My Farm on the Mississippi

Heinrich F. Hauser

MY FARM ON THE

MISSISSIPPI

The Story of a German

in Missouri, 1945–1948

HEINRICH HAUSER

Translated with an Introduction by Curt A. Poulton

University of Missouri Press
Columbia and London

Copyright © 2001 by
The Curators of the University of Missouri
University of Missouri Press, Columbia, Missouri 65201
Printed and bound in the United States of America
All rights reserved
5 4 3 2 1   05 04 03 02 01

Library of Congress Cataloging-in-Publication Data

Hauser, Heinrich, 1901–1955.
    [Meine Farm am Mississippi. English]
    My farm on the Mississippi : the story of a German in Missouri, 1945–1948 /
Heinrich Hauser ; translated with an introduction by Curt A. Poulton.
        p.   cm.
    ISBN 0-8262-1332-4 (alk. paper)
    1. Hauser, Heinrich, 1901–1955.   2. German Americans—Missouri—Perry
County—Biography.   3. Farmers—Missouri—Perry County—Biography.
4. Authors, German—Missouri—Perry County—Biography.   5. Authors,
German—20th century—Biography.   6. Perry County (Mo.)—Social life
and customs—20th century.   7. Mississippi River Valley—Social life and
customs—20th century.   8. Farm life—Missouri—Perry County.   9. Country
life—Missouri—Perry County.   10. Perry County (Mo.)—Biography.   I. Title.
F472.P4 H3813 2001   2001027270
977.8'69400431'0092—dc21   CIP

⊗™ This paper meets the requirements of the
American National Standard for Permanence of Paper
for Printed Library Materials, Z39.48, 1984.

Jacket Designer: Susan Ferber
Typesetter: BOOKCOMP, Inc.
Printer and binder: Thomson-Shore, Inc.
Typefaces: Futura and Palatino

Original German publication, *Meine Farm am Mississippi,* was by Safari-Verlag,
Berlin, 1950.

This translation is dedicated to the memory of

Heinrich F. Hauser

for his keen eye and great skill in describing
what he saw, what he felt, and what he sensed

and to

Helen Hauser and Huc Hauser

for their charity and dignity

Contents

Acknowledgments

Primary credit for this translation must obviously go to Heinrich Hauser. Without his superb narrative of a genuine geographical adventure, there would be nothing to translate. In my attempts to learn about the present state of Hauser's farm and its environs, however, I had a great deal of help from friendly and knowledgeable people, many of whom knew him and his family.

First, the support and approval of the two Hauser children has been invaluable. I have not had the pleasure of meeting Huc H. Hauser in person, but I have had a number of pleasant and illuminating telephone conversations with him. He very graciously gave me the collection of photographs taken at the farm and in the Altenburg and Wittenberg area that make up the bulk of the illustrations included in this volume. I also hope one day to meet Helen Hauser, a gracious lady, who, like her half-brother Huc, added much to my understanding of their father and gave me her blessing to pursue this translation and its very personal detail.

A number of Missourians in the region where Hauser had his farm were incredibly helpful, too. Boyd France and his sister, Jennifer France, are the present co-owners of the Hauser farm. Boyd granted me an interview and let me tour the farmstead. Their mother, Dorthy France, was raised for part of her life on the farm. She showed me the interior of the farmhouse and pointed out pieces of furniture made by Heinrich Hauser that still remain. Dorthy is a storehouse of local history and possesses a superb memory for names. She is currently making an inventory of books left in the farmhouse by Hauser, by other previous owners, and by her own family.

Henry Scholl, Dorthy France's brother, regaled me with stories about the period described in the book. He knew Heinrich Hauser and his wife, Rita, as well as Huc Hauser. Henry Scholl provided much new information for use in interpreting some of Hauser's descriptions. I refer to him numerous times in my notes, and for good reason.

There are others who have provided invaluable assistance simply by sharing their recollections with me in interviews. Harold and Rhonda France of Altenburg, Missouri, granted me a number of telephone interviews. Robert Fiehler, speaking with me at his automobile repair shop in Altenburg, was similarly helpful, as was his son, Gerard, whom I interviewed in an Altenburg café. Curt O. "Pat" Hoehne let me interview him at his home in Pocahontas, Missouri, and the Reverend James R. Marten, former curator of the Saxon Lutheran Memorial, provided a lecture and interview at the museum, located in Frohna, Missouri. George Thurm, in an interview at the Saxon Lutheran Memorial, also enlightened me with much historical detail about the initial Saxon Lutheran history in the area. I also received helpful information from Larry White, Perry County assessor, in Perryville, Missouri.

I am also grateful to Professor John M. Spalek of the State University of New York in Albany, who provided photocopies of Helen Adolf's biographical writings, and to the *Munziger Archiv* for sending me copies of its biographical entry for Hauser.

My Farm on the Mississippi

Translator's Introduction

In 1969 I found an intriguing volume on the foreign language shelves of the St. Paul, Minnesota, Public Library, called *Meine Farm am Mississippi*. I read the book haltingly, because my German was quite poor thirty years ago. As poorly and as slowly as I read it then, however, I was thrilled by Hauser's adventures and pleased to view his miniature portrait of what might be termed "postimmigrant America"—an accurate geographic description of an area settled by immigrants and compiled while some of the original settlers may have still been alive. Clearly, many of their children were alive and still practicing their Old World culture and language. In fact, Heinrich Hauser's decision to buy a farm in the area was clearly influenced by the existence of an exotic, anachronistic version of all that was familiar to him in his homeland.

I have since become a professional historical geographer, and I believe such portraits to be of immense value in the teaching of historical geography and of human geography. In 1996 I read the narrative again, this time with more facility, and found it even more powerful in its dual emphasis on historical geography and adventure. Upon reading the book this second time, I came to the conclusion that it should be accessible to American readers and thus should be translated into English. Further, it struck me as a fine idea that the field observations of a modern historical geographer be added to Hauser's postimmigrant study, contrasting his landscapes of fifty years ago with those of the present.

The unique perspective and detailed descriptions that Hauser offers act as a sort of time capsule for us, a view of a community that might otherwise fade from memory. Hauser died in 1955, and a great number of the people he knew and spoke of in this journal have also passed away. Except for this literary portrait, the piece of America about which he wrote virtually no longer exists. Only remnants of the communities he described remain today, and there are few remaining who witnessed the events of his time on his Missouri farm.

Curious to see what the region was like, in 1993 I visited the area where the farm is located, near the Missouri towns of Altenburg,

Frohna, and Wittenberg. I saw Wittenberg in its unfortunate last throes, inundated by the Mississippi for the last time. In 1997 and 1998 I made two more visits, to see what could be found of Hauser's landscapes and how they had changed in fifty years. Much is the same, but much more has been lost, I am afraid. Wittenberg simply capitulated to the Great Flood of 1993, and nothing that Hauser described of the town remains except its streets. It is interesting that his journal entries for the area begin with a severe flood in April 1945, with its attendant disruption of life and destruction of property, and go on to describe wildfires in the woods that disrupted and destroyed even more of what had already become important to him.

One purpose of this translation, therefore, is to provide a descriptive historical overview of that particular time and place that can be used to compare the landscapes of fifty years ago with those of the present, and to provide a historical geography of the region that might assist other scholars in their pursuit of local and American immigrant history.

But beyond the intellectual contribution of this translation, Hauser's journal is simply a *good read*. It is a true adventure story, written for a general audience. Although Hauser was a fine writer and is much revered in Germany and elsewhere, *My Farm on the Mississippi* is not a thick treatise or a work of high literature to be savored by the elite; it is a fast-paced, entertaining book, written to be enjoyed.

The Saxon Lutheran Villages

In 1838, 680 faithful Lutherans left their homeland in the former Duchy of Saxony, in what would later be the German Empire and eventually the Federal Republic of Germany of today. Being conservative Lutherans, they objected to the enforced liberalization of their faith by the imposition of a government-controlled (and taxed) church. Thus, they traveled to America to begin a new life in Missouri. Although their numbers had been diminished by the tragic loss of one of their ships and fifty-six of their members, they would become the founders of the Missouri Synod of the Lutheran Church, of Concordia Seminary in St. Louis, and of the "Log College" in their new town of Altenburg, Missouri.

These emigrants were not ignorant of the conditions they would face in Missouri. Although they were heading for a new world that they had not yet personally experienced, they had come to know Missouri vicariously through personal study and the advice of many others. They intended to settle specifically in Missouri because they had been well advised of the economic, social, legal, cultural, and natural conditions they would find there.

Gottfried Duden was perhaps the most influential of a great many writers who encouraged Germans to come to America. He traveled to the United States, and, in 1824, he bought land in Warren County, Missouri, at what would later be called Dutzow, a town on the Missouri River about forty miles west of St. Louis. Duden studied the area, and America as a whole, in order to provide Germans with factual, first-hand information. After three years he returned to Germany, where he published his *Bericht über eine Reise nach den westlichen Staaten Nordamerikas* (Report on a journey to the western states of North America) in 1829.

It is generally held that Duden, through the broad distribution of his work, was singularly influential in convincing his countrymen of the advantages of emigration from the overcrowded Old World, and of immigration to America, especially to Missouri. It is known through their own accounts that Duden had a great influence on the Saxon Lutherans in this regard.

In addition to the publications of Duden and others, personal letters by enthusiastic German settlers, writing to friends and relatives in the Old Country, were influential as well. These upbeat descriptions may have done as much as the published works in promoting "emigrant fever," as the great rush to America was sometimes called.

It became commonplace for groups such as the Saxon Lutherans and for organizations of young people simply hoping for a better life, to form *Auswanderungsgesellschaften* and *Siedlungsgesellschaften* (emigration societies and settlement societies), in which risks, funds, responsibilities, plans, leadership, and the like were shared, rather than borne singly. The prospect of undertaking such risky ventures as ocean voyages and settling in unknown territories became much less daunting when one shared the company of large numbers of fellow adventurers. Many Americans today are descended from members of such organizations.

The Saxon Lutherans initially arrived in St. Louis in the winter of 1838. In the spring of 1839 a large number of them traveled south on the Mississippi again and landed at the mouth of Brazeau Creek in Perry County, Missouri, between the towns of Ste. Genevieve and Cape Girardeau.

At nearly the same time, a group of German Roman Catholics from Württemberg, in southern Germany, embarked on a similar venture. They also headed for Missouri, and for the some of the same reasons. This group settled in Perry County at Apple Creek, and they are referred to locally as "die Badischen," a reference to their origins in what is today the German state of Baden-Württemberg. Some histories have stated that this group settled in Uniontown, Missouri, but local lore places them and their first church in Apple Creek.

The Saxons bought some forty-four hundred acres of land in Perry County, using their common fund of eighty-eight thousand gold *Thalers*. Here they established the colonies of Altenburg, Wittenberg, Niederfrone, Seelitz, Johannisberg, Stephansberg, and Dresden. Of these, Altenburg and Niederfrone remain and prosper, the latter having shortened and altered its name to "Frohna." Wittenberg, directly on the bank of the Mississippi River, also prospered, but it has since perished as a result of the capricious flooding of that river. Only one house in Wittenberg remains occupied today; all the others were razed after the disastrous 1993 flood.

Stephansberg was named for the Saxons' first leader, Pastor Martin Stephan. Unfortunately, Pastor Stephan became embroiled in a dispute in which his honesty and personal integrity were severely questioned. He was hounded out of the community and fled to the Illinois side of the river.

His replacement, Pastor Carl Ferdinand Wilhelm Walther, stood in dramatic contrast to the ousted leader. This well-loved man went on to lead his people in the events for which the Saxon Lutheran Colony and its offshoots in St. Louis are today justly well known. He was elected the first president of the Missouri Synod in 1847.

Thus, when Hauser arrived in Missouri, he found a thriving community of German Americans, proud of their Germanic culture and practicing it, despite the passage of generations. As with the immigrants before him, he treasured the familiar culture that greeted him in his new home.

Heinrich Hauser

Heinrich F. Hauser was born in Berlin on August 27, 1901, and died in Diessen, Ammersee (Bavaria), on March 25, 1955. During the latter years of the First World War he entered the German navy as a midshipman and began a lifelong love of the sea and seafaring. He traveled the world. In 1920 he published his first book, *Das Zwanzigste Jahr* (The twentieth year). After the war, he resumed his education and later became an editor for the Frankfurt newspaper *Frankfurter Zeitung*.

He gained a wide popularity with his first novel, *Brackwasser* (Brackish water), which won for him the 1929 Gerhart-Hauptmann Prize. His 1931 novel, *Die Letzten Segelschiffe* (The last sailing ships) has continued to find new readers up to the present. Following these two books, his literary output, in both fiction and nonfiction, was quite prolific, and he continued write prolifically until his death.

Hauser left Germany for America in 1939, pursued by the Hitler regime because of his anti-Nazi views and ostracized, ironically, by a public that had been confused by his dedication of a technical book on aviation to Hermann Göring, Hitler's minister of air forces and a top figure in the Nazi government.

Hauser first brought to America his Jewish wife, Ursula Hauser, née Bier, and two children, Helen (born in 1929) and Huc (born in 1933), each from separate marriages. Helen Adolf, a biographer of Hauser, described their relocation as being "like a cat carrying its kittens to a place safe from a threatening dog." Later, Hauser brought his companion, Olga Margarethe ("Rita") Laurösch, whom he would subsequently marry. He bought a farm in New York state, then sold it, and then he became a docent at the University of Chicago.[1] While in America he produced *Der Menschenleere Kontinent Australien* (The unpeopled continent of Australia) (1939), *Kanada* (1940), and *The German Talks Back* (1945). This last, pro-German work was remarkable for having been published in America at a time when this country was at war with Germany and had little sympathy for its people and culture.

1. Hauser said that he was a gardener and groundskeeper. Helen Adolf, his biographer, stated that he was a "docent" at the University of Chicago, but he himself mentioned nothing at all about academic activities there.

Growing weary of life in Chicago, Hauser moved to Missouri in 1945, and the events recounted in *My Farm on the Mississippi* occurred from then until he left the state as 1948 drew to a close, preparing to move back to Germany. His farm was located in the heart of a historically German immigrant community, and he felt a sense of belonging that he had felt nowhere else in America. The decision to leave was a painful one.

Hauser returned to Germany in May of 1949. He settled in Düsseldorf and renewed his writing career, which was chiefly involved with industry. His last published work, for the automobile firm Adam Opel, was *Dein Haus hat Räder* (Your house has wheels) (1952), an account of the ways automobiles reflect the comfort of modern living.

In 1955, only five years after his return to Germany, Hauser died, at the age of fifty-three, at his desk in his home in Diessen, Ammersee. He left behind a large body of original publications and numerous popular books, which were published in several languages and in several countries.

Translator's Note

A translator's job is not simply to convey words and concepts, but to be as true as possible to the original author's style and spirit. I was therefore faced with an immediate problem. Hauser had a fondness for dramatic, philosophical, run-on sentences; however, contemporary readers of American English would probably find the numerous dependent clauses more digestible if separated into smaller sentences. Yet the work abounds with colons and semicolons used in places where periods and new sentences would have been the better choice, at least in English. Thus, if I was to be faithful to Hauser's *style*, I would have to retain his lengthy and wearisome sentences and hope that English-speaking readers could plow through, forgiving me for what might indeed appear to be a bad translation. Meanwhile, if I was to be true to the *spirit* of what I believe Hauser intended, it would be necessary for me to reset the structure and syntax of the original. In other words, my challenge was to be as true as possible to Hauser's personality and character, as revealed in his text, and yet to make the work simple and readable, as I am sure he wanted it to be. If I have been successful in this, my translation will be a "good read" as well, not because of my skill, but because of Heinrich Hauser's style, spirit, and eye for detail.

Additionally, as a scholar, I am occasionally compelled to insert notes in places where Hauser's text can be supplemented with historical explanations and clarifications. Hopefully my comments will contribute to the reader's appreciation of Hauser's narrative.

It has been impossible, of course, to delve so deeply into the work of a famous writer without developing a sense of that writer's personal depth. Hauser's biographers have portrayed a most complex person, one who never seemed to be at peace with his own aims and convictions, both literary and personal. The muddle in which he found himself when he had dedicated a book on technology (in which he believed) to a member of the National Socialist Party (which he despised) is one example. His problematic relationships with the public and with his family are others.

As I became increasingly interested in his narrative, I found myself looking forward, with Hauser, to the arrival of his son, Huc, during each of the boy's summer vacations on the farm. In every reference to Huc, Hauser's love and respect for the boy fairly leaps off the page. And yet, as I have come to know Huc himself, I have learned that while there was indeed manifest love from his father, there were also harsh demands from the Prussian perfectionist, who carried his own father's rigid upper-class expectations with him throughout his life.

Huc Hauser kindly gave me, for use in illustrating this volume, a large number of black-and-white photographs taken on the farm during those summer vacations. He is prominent in most of these. Surprisingly, though, there is also a young woman present in a great many of the photos. This is Helen, Huc's half-sister. Helen was on the farm during at least one of her summer vacations, but she is not once mentioned by name in the narrative. She is finally alluded to in the last chapter, but only in a vague reflection on a troubled child in a painful relationship with her father, "caught up" in a "labyrinth of dispute and error"—another example of the problematic aspects of Hauser's relationships with his family.

"Henry" Hauser is remembered quite well by a number of people alive today in the region around the farm. He is recalled as "the peculiar man who sat in a tree house and wrote," and also as an outsider who had been welcomed into the community because of his own warm acceptance of its customs and standards. Hauser's character emerges as that of a very nice fellow in public who nevertheless had the capacity to be quite brutal, both emotionally and physically, in his private life. Helen Hauser told me that she believes her father inherited a tendency for clinical depression from his own father. She pointed out that Heinrich Hauser was married five times, and that the senior Hauser was married *seven* times, "the maximum number the law would allow." Helen thinks it possible that her father's multiple marriages were attempts to find someone to "rescue him."

This book is a powerful portrayal of the American farm ideal. As powerful as it may have been, however, Hauser's own Jeffersonian ideal of the perfection of self-sufficiency in an agrarian community and economy was flawed by his overestimation of his own capacities and those of his family and of his land. Of the 329 acres of land he owned, no more than 70 acres or so was in any sense true, arable farmland. He had been defeated by a farm in New York State that proved to be too small, too rocky, and too cold to be practically productive. Ultimately, the new and fecund farm in Missouri defeated him as well, precisely at the time that it had finally begun to be profitable.

As if the practical realities he eventually faced on the farm were not enough, his own inbred ideal of loyalty to family and community in his homeland—indeed, his *obligation* to his homeland and its ideals of family and community—so intruded on his conscience that he could not allow himself to accede to his own needs and desires. Thus the farm he so loved, this other, personal ideal, was forfeited to an older social ideal that he could not escape.

My Farm on the Mississippi

Introduction

It is easy to write about a great nation, even a whole continent, if you have been in the country only three to six months. In such cases you write of "first impressions," and if you have eyes in your head, you often create a surprisingly good illustration.

In this sense such pictures are like snapshots from a camera: They contain only the moment. They do not show the photographer's aspect, the thing caught in flight, or the photo's subject. They evoke the tension of the action but too often surrender to illusion.

It is much more difficult to write about a country in which you have spent at least some number of years. Then, you know much too much to surrender to impressions, but on the other hand you also do not anchor yourself in them. Troubled to get to the bottom of things, you conscientiously study them, and what comes of this, unfortunately, is all too often hard work.

So, I have totally rejected the impressions of my few first months as well as the experiences of my first few years in America. The cell cycle of the human body renews itself every seven years. It is therefore not surprising that a European generally sets roots in the United States of America only after seven such years. Since I only want to write about "my roots," I have limited this book to my last three years in America, the years of my second farm.

What I characterize here is not the typical America, although many Americans would like to believe that it is still typical. It is much more a remote and indeed dying America, and even in the German enclave I depict, it will no longer exist when the younger generation, whose members speak only English, has grown up.

To me this book is not about art or high style; rather, it is about life itself, about care and work, because life is so delicious for just those reasons.

Heinrich Hauser
Diessen-Ammersee, April 1950

11

Out of Chicago

In the spring of 1945 Rita and I were profoundly tired of our jobs. After her second winter as a temporary nurse at Chicago's Billings Hospital, I saw my wife becoming paler and thinner. She worked in the tuberculosis wing, and what she told me about it didn't please me. Toward the end of the war many safety measures against infection broke down, like the dishwashing machines, with the result that the dishes used by the tuberculosis sufferers were freely mixed with those of other patients. Nearby was the wing for the mentally ill, which, like all other departments, was chronically short on personnel. It was often the case that insane patients, among them very dangerous ones, wandered freely about the hospital. The thought that my tiny, delicate wife had the responsibility of escorting raving madmen to her duty station was not at all comfortable; even less so were the nearly nightly attacks on nurses on their way home.

Billings Hospital is in the large, parklike area of the University of Chicago. This "city within the city," with its eight thousand students, is enclosed on two sides by a heavily overpopulated colored district,[1] which generates enormous social pressure, not unmixed with racial hatred. For example, businesses and shops at the edge of the university pay doubly high insurance rates against disturbance and destruction than those that lie outside the racial boundary. In the last year of the war, the insecurity had grown so great that police patrol cars circled the huge hospital all night. Even so, because of the great number of exits and the dark parking lots all around, nurses on the night shift were so often attacked by muggers that it was usual for them never to carry more than a dollar or so on their persons.

Whenever I could, I always picked up Rita myself around midnight. But that was not always possible, because from time to time I also worked the night shift in the warehouse at Marshall Field's. By day I was a gardener at the university. "Gardener" sounds nice, and there was always something to do in the constant planting

---

1. Hauser referred to African Americans as "colored people" or "negroes." I have not changed his choice of words in favor of modern usage.

of flower beds and replanting of shrubs. In the main, though, we gardeners were really glorified street sweepers, because it was our duty to clear the huge lawns of what those eight thousand students threw away, and that was quite a bit. The greatest number of empty bottles, to speak of the most prosaic of the litter, we found under the dormitory windows of the coeds. Most of us "gardeners" were older men, and we had become somewhat cynical through our life experiences; our commentaries about students at lunchtime were accordingly somewhat biting. "The younger generation," said the old Scottish head gardener, "really can't read and write anymore. If this spoiling of the youth goes any further they'll forget how to eat by themselves. Then the professors will have to feed them their porridge with a spoon, just like the so-called knowledge of today, and they'll have to be careful that none of them starve."

The garden work brought in thirty-five dollars a week. That was the reason that five evenings a week I also spent five hours, until midnight, working at Marshall Field's in the "Special Service" department. Special service consisted of providing each sales counter with cartons of goods for the next day's sales. The cardboard boxes were delivered by truck, bundled together, and dragged with steel hooks into the elevators and then brought by little roller platforms on each floor to the proper station. The contrast between the wind blasts outside on the loading platform and the used, overheated air inside was tremendous, as was the difference between the impoverishment implied in this work and the splendor of the goods displayed. Even though everybody knew that the detectives patrolled every floor at night—big, friendly, but sharp-eyed fellows mostly—for many of us there was still temptation. Thus, four months before Christmas, the "Special Service" began with ninety men, but by Christmas the number had fallen to nine. Eighty-one men had been fired for dishonesty, but the nine that remained did as much as the ninety at the beginning and had good prospects of being hired again. In such work honesty was at a premium. The work was not uninteresting, especially since the warehouse was often open until ten in the evening, and we worked among the customers. Here again we experienced high emotion in the tension between the races. In its personnel policy Marshall Field's took care to be thoroughly neutral. White and colored were hired in exactly the same racial proportion as that of all Chicago. But Marshall Field's did not sell to colored people, and since the doormen were not always able to keep the coloreds out—because they didn't want to create a commotion—it often led to embarrassing scenes. For instance, when a wealthy negro lady demanded to be served and was refused, she scolded even more loudly that the goods were cheap and shoddy . . . until she had to be more or less tactfully removed. It was especially difficult when a colored lady customer had to be turned down for service by a member of her own race.

Around ten o'clock our group went to the employees' cafeteria on the top floor for ten minutes, where there was free coffee for us. On more than one occasion I was handed an anti-Semitic or other race-hate pamphlet under the table by someone unknown. Chicago is a battlefield of unamalgamated races: Radio stations, with their polyglot programs in Czech, Yiddish, and Italian, reflect this in large scale, and Marshall Field's, with its dialect-speaking charwomen, its Viennese interior decorators, and its cute women's style experts from Berlin, reflect it in the small.

The most interesting nights were the Sundays before Christmas. The really rich didn't shop only during business hours; these privileged few were invited to a private showing and secluded shopping, which Marshall Field's dedicated to them on Sunday by special appointment. Only a small elite among the many thousands of personnel were present then—chiefly charming young women who served as guides, since most of these customers were men. It is difficult to buy really expensive things when your family has everything; but these young women showed great expertise in leading their customers over the slippery slopes of luxury. The rich felt thankful for the lively interest the pretty guides took in their families, and they were enthusiastic about the intuition with which the young women speculated about family members' possible wishes. All of a sudden the patron had bought double or triple the amount he had planned, and sometimes it was not only the beauty and cleverness of the young woman that led him to do this, but also the champagne and other creature comforts that Marshall Field's so good-heartedly and freely placed at his reach. Last and finally came the great moment when the customer plucked out his checkbook, and this was the moment for which Marshall Field's was impeccably ready. Discreetly in the background followed a department head, and behind him an adjutant carrying a tray with a dozen ready-to-write fountain pens, each according to taste; then the signature was made, and on such Sundays single checks of fifty thousand dollars were no rarity.

As interesting as this was at the time, it was equally exhausting. Rita and I both felt consumed by the huge city. Admittedly, our apartment in Drexel Square in south Chicago wasn't bad, considering our income. We could see greenery outside, and we were near Lake Michigan, which offered wide horizons of freedom. In our neighborhood there were also shops of all nationalities, where we could buy a good loaf of the European rye bread that we so hungered for. But when I observed the children at their unconsciously desperate favorite play—smashing glass bottles, and firing toy pistols at passers-by, whom they unconsciously hated—I had to think of my own children at school in New York state, and knew that it was high time to give them a real home. From our salaries and from the proceeds from earlier books we had saved perhaps three thousand dollars; something ought to come of that. We didn't want to return to the north under any conditions; six icy winters there, with temperatures down to forty below, had been enough for us. In every free moment we put our heads together over maps. We knew the West Coast well, and it enticed us most powerfully, but it was too far away; the trip would have wasted too much money. The second greatest magnet was the Mississippi. A decade and a half before, I had traveled the river from New Orleans to Minneapolis, practically its entire navigable length. It was an unforgettable experience, a weeklong voyage on one of the last of the old paddle wheelers, the *Tennessee Belle*. Now the river lured me back forcefully, especially when we discovered a chain of small cities with romantic French names, like Ste. Genevieve and Cape Girardeau.

The strongest force that drove us out of Chicago was the press and the radio. In the city we couldn't escape either of these, and in the spring of 1945 both of these outdid themselves with reports about the inundation of German cities under the rain of bombs, about the destruction of everything in them that we had once loved. I couldn't bear it anymore. Even the short-subject news in the weekend movies were full of it. For the first time in my life I wrote poems . . . bad poems and in English. They were about the cities in which my

"Perfidio" in front of the
Hausers' Drexel Square
apartment on Chicago's
South Side.
Photo courtesy Huc Hauser.

children were born and which they would never see again, and
since these poems were dedicated only to the children, they had
no influence on public opinion at all.

One thing more remained unspoken between Rita and I: In order
to break free of Chicago, in order to begin the search for a farm, we
needed a car. In the past, when I had bought a car, I paid much more
than I could afford. I knew that Rita mistrusted me greatly in this
respect, and rightly so. In addition, the supply of cars in these last
war years was quite meager. New cars, when they were available
at all, brought high premiums, and even used cars were overvalued
because of scarcity. So it happened that one evening I came home
with a very red face, and Rita knew instantly:

"You bought a car!"

"Yes, we have a car. It has a name, too: It's called Perfidio."

"Why is it called Perfidio? That's not a kind of car, is it?"

"No, but when I tried out the radio, the first song I heard was 'Perfidia,' and that's what I called it."

"It will do its name honor," said Rita, bitterly.

"It will not. It's a Packard, really old, a 1928 model; it's a sedan. It belonged to an old lady and it was garaged for nine years without being used. It's like new."

"Like new is good. What did you pay for it?"

"Four hundred fifty dollars. The salesman wanted six hundred. I talked him down, and I'm proud of myself."

Rita sighed, "That speaks to the need. Where is it?"

"Downstairs. In front of the door."

Between all the modern cars, the old sedan contrasted like an elephant among sharks; its enormous headlights were better compared to old coach lanterns, but its paint and brass twinkled wonderfully in the streetlights.

I started the powerful machine, and as old as it was, the 120 horses growled like a lion after a rich feast.

"The die is cast. Tomorrow we leave."

Toward the South

The divided highway from Chicago to St. Louis is not distinguished by natural beauty. The flatland is only admirable for its abundance—cornfields spread as far as the eye can see—and if the farmhouses are big and radiant in their white paint, the barns and outbuildings more resemble railroad station halls, and when the sun shines, their shiny metal roofs blind the eye. Every farm has signs that tell which variety of corn has been planted here, and with which chemical fertilizer the string-straight fields have been treated. Passing by later in the year through the six- to nine-foot-high forests of dark green corn, every driver can immediately compare the seed and the quality of the fertilizer by the condition of the crop; the signs the farm supply merchants have placed there act as advertisements. And as if the Lord of Nature had not done enough by creating this rich land, he had deposited oil and coal under it. On long stretches you encounter series of small mines, surrounded by trucks, immediately below their head frames, from which the black blessings cascade, while on other stretches oil pumps stand about on the land, with long arms bobbing slowly up and down, all the machinery fully automatic, without a single soul far and wide. And flames flare from vent pipes thrust up out of the cornfields by the hundreds, burning off the unwanted gas. By day you can barely see their flames, but by night these giant flares paint a magical portrait beside the highway.

We began the six-hundred-mile trip at noon, in the expectation that we could complete the greatest part of the stretch at night, because the traffic would be less heavy.[1] That was a mistake, however; the deeper it darkened into night, the heavier the truck traffic became, and the faster their speeds. After I had gotten used to Perfidio and found the machine sound in all parts, and reliable, I drove at an even speed of fifty-five miles per hour; nevertheless, more and more of

---

1. The distance from Chicago to St. Louis is actually only about 325 miles. Hauser may have meant to say "kilometers" here, so that his German readers would better understand the distances involved. Throughout this book, however, Hauser generally overstates distances.

the powerful diesel trucks with their enormous aluminum trailers thundered by me, not to mention buses, which travel at a good sixty miles per hour and leave Chicago on their night schedule every hour for St. Louis.[2]

After a few hours, the endless rising suns of oncoming headlights and the constant watching for the bright strings of lights that show the contours of the trucks became rather stressful. Shortly after midnight we drove into one of the many "motels," which announce themselves with equally bright chains of lights from far off. Motels are auto courts at the edge of the highway, with a restaurant, and a number of wooden cottages in the background, each with an open roof covering a spot under which you can drive your car. You step out of your car directly into your bedroom, and this is much more comfortable than in a hotel. This night just before spring was very cold, and our motel cabin had no heat. It had an automatic oil-burning heater into which you dropped a quarter, the equivalent of a German one-Mark coin, and this made the atmosphere very cozy within ten minutes. In the interim we went to the restaurant, where the short-order night cook prepared simple dishes like bacon and eggs in minutes, and where the bar was open all night.

The next morning my first concern was naturally Perfidio: with the greatest of relief I determined that the machine had used practically no oil, and no water at all, a sign that the motor was sufficiently cooled, and that the fuel consumption had been held to acceptable limits. The only thing that did not please me were the tires. The old, long-legged Packard had eight-ply truck tires with excellent tread, but they were a good ten years old, their rubber already visibly cracked, and at high speed or with sharp braking, I feared a blowout. In the meantime, Rita bought a thermos bottle and we had it filled in the restaurant. Much happier, because we were free of worry, we began the second day of our trip.

We reached St. Louis shortly before noon, but we practically didn't stop at all. Even with a completely familiar car, driving through a strange American city requires such concentration on road signs

2. To preserve tires and fuel during World War II, a national speed limit of thirty-five miles per hour was established by a War Order. It is unlikely the order had been rescinded by this time.

and traffic that sightseeing is almost impossible. Here we found not only that St. Louis has a vast, thick network of streetcars, but also that Perfidio could not keep up with modern cars in acceleration and braking. So we parked for just ten minutes under the huge Mississippi River bridge; we couldn't drive quite so fast past our first view of the river that had been our main goal.[3] There was high water, as there almost always is in April; I greeted the muddy waves like an old friend, but I was sad about the changes time had brought to the view offered. There, where fifteen years before the old paddle wheeler had tied up, lay a streamlined colossus of a motor ship called the *Admiral*.[4] But it didn't cruise to New Orleans anymore; instead, it is by day an excursion boat and by night a floating dance palace. The old steamboats lay on the other side of the bridge, or, more accurately, what was left of them. Steam no longer rose into the air from their crowned double smokestacks. Decks sagged, graying paint peeled, and the ornate fretsawn woodwork of the galleries hung in tatters.[5] Only one single steamboat still apparently had a life as a showboat, or theater ship; but they had torn the machinery out of it, so it had to be towed.[6] So we threaded our

3. This is the Eads Bridge, designed in 1874 by James B. Eads. Though now closed to automobiles, the bridge has been preserved for its historical and architectural value and beauty.

4. The *Admiral* is today the St. Louis President Casino, tied to the same spot on the St. Louis Levee. Its machinery has been removed, so it no longer cruises the river, but it continues to serve as a floating nightclub and gambling hall.

5. The St. Louis Levee in the 1940s was the final home of a number of the older steamboats Hauser mentioned. Their time of service was long past, and they were indeed neglected and in bad repair. One evening at about the time of this narrative the river's level dropped quickly. Tied close to the shore, some of the boats were beached on the levee when the water receded. There was no way to refloat them until the river rose again—but by then, the boats' planks had dried and shrunk. When the water rose, the boats did not.

6. This was Captain J. W. Menke's *Goldenrod* showboat. When Hauser saw it, it was still afloat on its own hull and had been moored at the levee, powerless, since 1937. Absent the weight of its boilers, engines, and paddle wheel, its hull had acquired a curious wavy irregularity along the deck and throughout. By 1955 its superstructure had been removed from the

way back into the stream of traffic, southward, following highway number 21.[7]

In many places, Highway 21 crossed relict channels of the Mississippi; for seconds, then, you could get a fleeting view of the dawning of the "bayou" world: dark ribbons of water, overhung with giant cottonwood trees with silvery, shivering leaves; flatboats with negro fishermen; dilapidated little houses, balancing on high piles over thick willow groves and seeming to float by through the sunspots and the deep shadows—the whole thing appearing mysterious and dim. Right at the edge of the metropolis was an absolutely foreign and powerful world. After about fifty miles, the road snaked up a limestone bluff and thirty miles later struck the river once more. There lay the first of those places that had sounded so enticing on the map: Ste. Genevieve. With its seven thousand residents or so, the little city could have just as well been in Europe, since it is more French than American in character. It is a Catholic center, and housed under the wings of its cathedral were a bishop, a seminary, and a nuns' cloister. We were tempted to stay here and speak with the white-cowled nuns, of whom it appeared that many were German, judging from the China-babble of languages that dumbfounded us here; yet the day was still young, the city was not directly on the river, and I had a feeling that here was only the beginning of that world that so oddly and magnetically drew me. We directed our course toward Cape Girardeau.

Between St. Mary, the next town, and the "Cape," as it is known for short, the road covers thirty-six miles along a ridge and doesn't touch the river again. Without knowing why, I drove slower and slower and observed the landscape ever more attentively, until finally I stopped the car altogether at the top of a rise. To the east, approaching the river, wooded hills sank away, row after row, and wherever the slopes were gentler, there were fields spread out halfway up, and in

undulating old hull and placed on a modern steel barge. The *Goldenrod* still entertains audiences with old-time melodrama and olios, but does so today at St. Charles, Missouri.

7. What Hauser referred to as Highway 21 in 1945 is today U.S. Highway 61. U.S.G.S. topographical maps published in 1947 show it as Highway 25.

the valley floors, clean white villages were grouped around spiked church steeples. Near and far, as far as the eye could reach, chicken hawks circled in the sky, and as I checked the treeline I found the majority of the stands of trees to be oak, beech, and pine.

"Is there anything about this countryside that catches your interest?" I asked, finally.

"It could be anywhere in Germany," Rita announced; "Harz, Taunus, Westerwald, Erzgebirge—there is something here from all of them."

Perfidio had hardly reached a mild thirty-mile-per-hour speed when I stepped on the brake. A side road cut across 21 at a right angle and on a road sign stood the unambiguous name "Stuttgart." Another six miles, and yet another crossroad pointed toward the Mississippi and "Dresden"; right after this a wider road led to Altenburg and Wittenberg. There was no more doubt: Right here in the middle of a French enclave there was apparently a German province, but the places were so small that they did not appear on the maps.

I would have preferred to drive down to the river on one of those side roads at that instant, but the surprise was so great, and the day so far gone, that I wanted to sleep on this proposition, especially since opportunities for overnight accommodations in these little villages did not appear very favorable.

In many steep waves over hill and dale, winding along rose-lined field borders, 21 brought us to Cape Girardeau near evening. Below, at the river, at the foot of this hilly city, there was much activity. In the last ten hours the Mississippi had risen nearly ten feet, and all of the railroad yards were already flooded. We arrived at exactly the right time to see a train on the Frisco line steam in; the locomotive pushed a regular bow wave before it, the wheels of the rail cars disappeared under the floodwater to above their axles; a long wake followed the train, which appeared almost to be floating. Since nobody could get out of the cars at the regular station, the engine drove on to the freight station, which was somewhat higher, where the passengers walked dry-shod across a plank bridge to their waiting cars. Shortly after this, a switch engine with a traveling construction crane rolled in, and a column of negroes began to build a raised temporary track. This transpired as simply as this: The crane laid whole sections of mounted rails, complete with their ties, on top of the sunken tracks.

Fishermen, who had tied up their boats on the main street, appeared not to expect much for all this. "Farther south," they said, "there are a lot of railroad bridges under water. They're threatened with being washed out, and this is probably the last train."

We had had no concept that the river could rise with such breathtaking speed; we wanted to settle on its banks, and the spring flood moved us to quite new respect. Now, in order to see how it looked from the other, flatter bank, we drove over the Mississippi bridge, even though the bridge attendant assured us that the toll would be wasted, because we wouldn't get far. The man knew what he was talking about: Less than a mile from the eastern bridgehead, the road disappeared in a vast lake of floodwater, but it was still wonderful to experience the mood of the deluge in the light of the setting sun, to observe flocks of white herons wading, the only creatures happy in their new paradise.

Cape Girardeau is probably the first truly southern city on the Mississippi's course. Many of the pink- and red-painted houses have cast-iron balconies running their full width, supported on iron columns over the sidewalk below, so that one walks in the shade of these galleries. In the midday hours the green-painted shutters on the windows are kept rigorously closed, and in the "plaza," high above the river, stands a bandstand, around which one promenades in the evening. Distinguishing its southern sympathies, there is also a memorial to Confederate soldiers in this square, the first memorial to the defeated side of the Civil War that I had seen. The bronze warrior looks a little like Napoleon III; his expression and his weather-worn coat mirrors something of the deep bitterness this most harsh of all American wars had left in the hearts of the South even until now.

As we lay down to sleep under the humming overhead fan in our hotel room, I took Rita's hand:

"Tomorrow, I think, will be a decisive day. We'll drive back to the road with the German names and follow it to the river. I believe that we will find a home there."

The Doctor's Little House

The next morning I visited a doctor; there was an infected finger to repair, and the previous evening I had seen a sign in the main business street with the not unfamiliar name "Dr. Schultz," which had inspired my confidence. Dr. Schultz turned out to be a man well into his seventies who didn't take long to begin to mention his reminiscences. Yes, he himself was born in a nearby German locality that I would want to visit. The region had been settled by Saxon Lutherans, who had emigrated a hundred years earlier for religious reasons, and at approximately the same time a group of Württemberg Catholics had come to the same area. All good people, who had retained their customs remarkably well—quite aside from their religious convictions—because the region was so secluded from the rest of the world. His parents had told him of the famine the settlers had endured in the first winter, and that in his own youth there had been many fever epidemics.

He hauled out a packet of somewhat gruesome photos of surgical operations: "This was my first work here; I was the first real surgeon at the Cape. At that time there wasn't any running water—I had to operate using boiled Mississippi water. We've only had a really modern hospital for ten years, which the Rockefeller Institute helped us to build, but I regard it as my true life's work. Yes, there are a large number of farms for sale; the young people all want to go to the city—community doesn't survive any longer; it's a pity. But in your place I'd be careful: First, let a farm agent test the soil, and, above all, don't let anybody know you're after a farm—that drives the prices up. The best thing to do is to stay in the area a couple of weeks. A roof over your heads? That shouldn't be difficult. It just occurred to me: Dr. Moran, an old colleague of mine, built a little house at Tower Rock in his last years, right on the river.[1] He died last

---

1. Dr. Schultz may have disremembered the name of his colleague, or Hauser may have recorded it incorrectly. The doctor who built the retirement house was named Van Note. It is possible that Hauser changed the doctor's name out of respect. George Thurm, at the Saxon Lutheran Memorial, said

year, heart attack—a nice sort of death. He didn't have any relatives, and as far as I know, the little house is empty."

I don't believe that such things are coincidences. Here was, I felt, a call from destiny, and I didn't delay for a second to follow him up: "Who's managing the house?"

"Wait, now; that may be Charlie Bussen, a quarry owner; he bought the land around Tower Rock at the outbreak of the war to ship limestone in barges. There is a sort of a harbor there. But that was a failure; as far as I know, the business is idle."

"Where does Charlie Bussen live?"

"Right down on the river."

I found a charming house, overgrown with wisteria, and inside a tall man with black hair and sparkling white teeth, who shook my hand cordially: "The doctor's house? I bought it last year; there was a good shed with it that we could use to store our machines. You want to rent it for the summer? . . . Sure, why not; not a bad place for a writer. Only you'll have to have the cistern checked; it's been unused for so long I wouldn't trust the drinking water. The rent?" He laughed out loud: "I really don't know about that; I have a half dozen houses like that one along the Mississippi, and most of them I let people live in for free. Well, if you absolutely have to pay, let's say five dollars a month."

I laid the first month's rent on the table, and we sealed the agreement with a handshake. Nothing more was necessary for Charlie Bussen, that I knew. What I didn't know was that this man in khaki shirt and pants, who looked like any other laborer, was in reality one of the richest people in the whole region.

"Have you finished packing?" I asked Rita in the hotel. "We have a house. It's two miles south of Wittenberg on the Mississippi. Come on, let's go."

She shook her head: "What kind of a house? How—why? You haven't even seen it. In God's name, you didn't buy it?"

---

it seemed that Dr. Van Note had lost his license to practice medicine in Massachusetts because of morphine addiction. Van Note perhaps sought out this quiet backwater as a place safe to continue his practice. Thurm also stated firmly that Dr. Van Note was "a wonderful doctor, and, in fact, he brought me into the world."

"No, no, we can only lose five dollars in the worst case. It's a long story; I'll tell you on the way." At that, I picked up my shaving gear, and five minutes later we were in the car.

From the crossroads where 21 cut across the gravel-strewn road to Altenburg and Wittenberg, the landscape grew prettier and prettier. It curved through the wooded hills; the young foliage gleamed, illuminated by the sun, and we saw many things that we had not encountered since Germany. For example, there were women working in the fields; they had to be Germans, since no American women would work in the fields here. They wore big straw hats or white bonnets, and as they looked up from their hoes, they waved to us, as if cars were a rarity in these parts. We met schoolchildren, on their way home for lunch; some were barefoot, and even this was astounding, because elsewhere schoolchildren were almost always transported by bus. Tractors puffed, covered over with big white parasols, even though it was not really hot. But the main impression was not the modern mechanization, but the signs of an earlier time. Farmers drove in the direction of the main road in little high-wheeled horse-drawn wagons that in Germany would be called *Sandschneider* (sandcutters)—two-seaters with a bonnetlike leatherette cover over them. Mule teams hauled golden ears of corn to market, and when an ancient Ford of the fabled Model T wobbled by, even it appeared to belong to an earlier age, like the horses.

After about eighteen miles we reached Altenburg, with "Population 298" proudly proclaimed at its entrance. Despite its villagelike size it gave the impression of a small city: There were many garages, two churches, two stately stores, a cubic, tile-red little "Bank of Altenburg," with posts in front on which to tether horses. The place had a little park as well, and under high maple trees stood a tiny log cabin, its joints white limed, with the inscription: "This was the first Lutheran seminary west of the Mississippi, founded in 1839."

Nearby, and very reassuring, to my way of thinking, lived a Dr. Fischer, so there was a dentist in the place as well.[2] Words cannot describe how enormously homelike this idyll was for us after the years in Chicago.

2. Dr. Theodore Fischer was a medical doctor; the dentist was Dr. Edward Lottes.

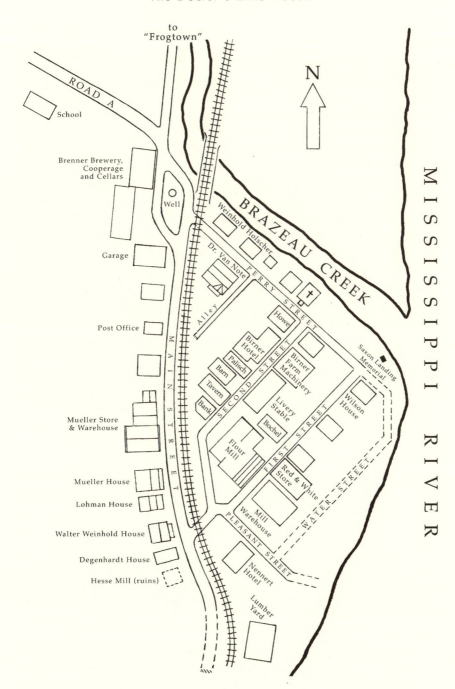

Wittenberg, Missouri, in about 1945. Adapted from a sketch by George
Thurm, of the Saxon Lutheran Memorial, Altenburg, Mo., May, 1998.

"Are you thinking about staying in the fabulous little house overnight?" asked Rita.

"Of course; dead or alive."

"In that case, we ought to buy a few things," she suggested. "Tomorrow is Sunday."

It was remarkably perplexing to enter the store, ask for things in English, and then receive the answer in the best German. At the bank, where I cashed a couple of traveler's checks, the customer before me said, "Na, Euscheen, gib mir mal heite hunnerd Dollaars."[3] Not only did everyone speak German, but they all said *du* to one another.[4] It was like a fairy tale—it was America, and yet it was not America; time and distance were magically crazy.

From Altenburg the country road descended fairly steeply into the Mississippi Valley. Run-over rabbits attested to the richness of the wildlife in the area; a huge hawk flew up just before our wheels, and partridges ran zigzag courses in front of us, then whizzed off in low flight into the undergrowth; the road plunged visibly into a wilder land. There was one more chain of hills to cross, and then, just before a curve, we met a sign: "Wittenberg, Population 89," and a minute later we stopped at the river . . . in a miniature Venice. There, in an irregular and broad pattern, stood a couple of dozen white wooden houses with corrugated metal roofs, their first floors in water up to their windows. Some had boats tied to them, and from others wobbly wooden bridges led toward higher land. Those houses standing in the deepest water seemed completely abandoned; half of Wittenberg evacuates its homes when the great spring flood comes.[5] Clinging to

3. "Well, Oyshayn [Eugen, or Eugene; in proper German, pronounced 'Oy-gain'], gimme a hunnert dollers t'day." Hauser was relaying the dialect he heard, which was softer and more gently personal than the correct German he wrote.

4. Hauser's term was *duzten*. Germans in their homeland usually speak formally to one another in public. It is normal to say *Sie* (you) along with the person's surname and title: *Sind Sie Herr Schmidt?* (Are you Mr. Schmidt?). *Du* (thou) and the given name are used only in the familiar, as among family members or close friends, as in: *Bist du wohl heut', Karl?* (Art thou well t'day, Karl?). The word *duzen* (pronounced "dootsen") refers to speaking in the familiar by saying *du* (thou).

5. The Great Flood of 1993 was the end of Wittenberg. By 1997 only those houses on the road leading down into the riverbank town appeared to be

The last post office in Wittenberg. One of the few Wittenberg structures remaining intact, the building is now in use as a souvenir and gift shop. *Photo by Curt Poulton.*

the edge of the clay-yellow flood ran the rails of the Frisco line;[6] to the left stood a small station building, and on the right was a single row of houses in a line, about twelve in all, Wittenberg's Main Street. The first of these was a tiny corrugated sheet-metal post office, and just about the last was Müller's Store, which would play a large role later in our life. On the other side of the tracks, in the water but still reachable, there was a tavern, and we directed our steps into it first of all. The bar was crowded with railroad workers, each of whom

---

occupied. Only three houses remained on the river's shore in 1993, and what was left of the town looked exactly as Hauser described it in the 1945 flood. Just a few months later, the last house would be dismantled. One structure, the old Brennert Brewery, remains alongside the bluff at the floodplain's margin. It is in use as a residence. All the original wooden houses and other buildings have disappeared as if they had never existed.

6. The St. Louis and San Francisco Railroad. The line is currently labeled on topographic maps as the Burlington Northern Railroad and would today be the BNSF (Burlington Northern Santa Fe).

had a bottle of Falstaff or Budweiser in front of him . . . and no glass. For German tastes, American beer is somewhat sweet; perhaps that was the reason why there were a great number of saltshakers on the bar, from which everybody salted his beer to his own taste. We had really only come in to ask the way, but the tavern droned with voices that spoke only of astonishment at the flood, and since there were a number of stuffed animals, snakeskins, and giant turtle shells to admire, we did as the natives did and drank our beer from the bottle, with salt. The main attraction of the place seemed to be—next to the beer—two pretty and prominently voluptuous daughters of the proprietor.[7]

When I was finally able to drown out the commotion over stuck trains and flooded houses, an old fisherman in high rubber waders spat out vigorously:

"Tower Rock? Don't think you can get through. What do you think, Bill, how do you reckon the Dip?"

"Oh, about two feet. But the gent has an old high car, I saw it already. Could be that he'll get through all right."

Bill, also in waders, was a negro with a piratical appearance; you could easily be afraid of him. We had no idea at the time that he and his companion, Leo Harris, who looked just as threatening, would one day count among our best friends.

Armed with good counsel about how "the Dip" was best over-come (why it was so called I didn't understand), we started on our way. A half mile outside Wittenberg, a small road bore southward through the woods—this must be it. The entrance went through a brook; immediately beyond, it rose steeply into the forest; I shifted to first gear and gave it some gas . . .

I am an old-time driver and until then thought that there were no new tricks for me to learn, but this "road," mildly speaking, was a surprise. Perfidio, with all 120 horses, tumbled into deep, washed-out ruts, skidded in sticky clay, and, strewing a cannonade of rocks

---

7. The tavern was Nick's Place, owned by Nick Lungwitz, who had purchased it from a longtime owner named Eisenberg. Upon Nick's death, his wife, Bertha, operated the tavern as "Bert's Place." The two voluptuous daughters were Pearline, who died young, and Delores, who is the wife of Henry Scholl.

Tower Rock, 1998. *Photo by Curt Poulton.*

behind it, barely reached the first hilltop. I estimate the slope at thirty degrees.[8]

Then we went up and down through a tangle of hills such as we had never experienced, through potholes half a wheel deep, between boulders and fallen trees that grabbed at the fenders on both sides, and never ever came to an end, so that six miles felt like sixty. Rita clung to the assist strap of the old-fashioned coach and asked from time to time, "can't you drive a little slower?"—but I couldn't, or we would have gotten stuck. At last the track of the road went through a deep ravine down to the river, and right there, where it opened onto the railroad embankment, lay the notorious Dip. It was quite simply the borrow ditch beside the embankment, normally dry, but now filled by the floodwaters.

I didn't dare to drive straight into it. I had no rubber boots, either, so there was nothing else to do but to roll up my pants and

8. This road is indeed rough, unpaved, and steep in places. Only the first quarter mile or so is really so daunting, but it is nowhere near 30 degrees from level—it is more like a 15 percent grade.

Tower Rock, 1946.
Photo courtesy Huc Hauser.

wade through. The water reached all the way up to my thighs, and the worst of it was that the bottom consisted of soft, black mud. I carefully took off the fan belt, told Rita to raise her legs high, and gave Perfidio the gas. A mighty cloud of steam rose from the boiling hot radiator, and then a gush of water broke through the floorboards. The rear wheels milled dangerously, while the car moved forward only inchwise. Then the radiator heaved steeply upward, and with a desperate leap Perfidio made it across the tracks . . . we felt ourselves rescued.

From the height of the railroad bed we looked across the river . . . but the other side could barely be seen. The level land on the other side was flooded four miles inland, and even on our side the poplars stood deep in the water. The view was breathtaking in its magnificence; as far as we could see, great uprooted trees drove by, turning around and around themselves, so that the crows that floated along with them kept flapping up. Breaking through heavy clouds, the sinking sun reflected off the yellow waves eerily, and the nearly drowned mass of the undergrowth of willows beat wildly about in the suction of the waves. A half mile south, a gigantic gray cone of rock loomed out of the river; that must be Tower Rock, and from there came the powerful sound of rushing water.

The way in, between the railroad and the rocky bank, was relatively simple. We crossed the tracks a second time, and from there the road wound itself into the woods again, but in front of it lay a triangular piece of open land; right in the middle of the triangle was a garden with blooming pear and peach trees, and in the center of the garden stood a small, brown, wooden house: the doctor's house.

Rabbits scrambled away as Perfidio broke through the high grass of the entrance, and then we were stopped in front of the little house, still trembling from the excitement behind us, and trembling in advance of what lay before us. Like children, we held hands . . . it was indeed a fairy tale, the whole of it; such things couldn't really be true. And yet it was true; the noise of the rushing water wrapped around us like a wave.

I stuck the key that Charlie Bussen had given me into the little padlock, and the door opened with a creak. There to the right was a little kitchen with built-in shelves, a hand-pump, and a small iron stove. To the left was a small room that held nothing besides a couple of empty bottles. Ahead, toward the flooded river, we entered into the only large room, apparently a living room, with windows on three sides and a tiled fireplace with a puddle in front of it left from the last rain. The remainder of this room was also completely empty except for an iron bedstead that leaned against the wall—and a big bundle that hung from the ceiling by wires. On closer examination, this was found to be a rolled-up mattress that someone had tried to protect from the rats in this manner. Naturally, everything was covered in thick dust and coated with spiderwebs.

Perhaps it is good that people are not allowed to dwell too long in fantasies. Evening fell and promised not more than an hour's daylight; there were naturally a tremendous number of things to do. As the luggage was dumped in a disorderly circle around Perfidio, I didn't fathom how I could bring it all in: cots, sleeping bags, and

Doctor Van Note's "little house." Huc is holding Holla.
Photo courtesy Huc Hauser.

a complete cooking outfit, quite aside from all the other stuff. In this wilderness, all the electrical apparatus was ridiculous, and we ignored it.

"Let me take care of the house," said Rita; "the most important thing, first, is water."

I left her to her broom and brushes and checked the cistern. As I suspected, the lid did not close tightly, and the hand pump was dried out and rusted. The first bucket, let down on a rope, brought a live tarantula and a dead bird to light; both displeased me, for there was nothing for us to do but manage temporarily with boiled Mississippi water. The path to the river led through the neglected garden, crossed the railroad, and then appeared to lead to a big peninsula. That had become an island, however, because of the flood, so I went farther down the path to the water and scooped the bucket full. One of the large cans in the kitchen turned out to be a clean flour container, and after six times back and forth Rita was temporarily provided with water. In the meantime, all the windows stood open, the stale air was withdrawn, and on top of the little woodstove our Primus stove hissed.

When, after several collapses, the iron bedstead finally remained standing, the mattress was beaten and a keg was set up for a table; when the first tea steamed in the tin cups; and as light shone from the first candle, then the fantasy took us into its spell. Looking into each others' eyes, we wordlessly shook our heads: How on earth had all this come about? We didn't know ourselves. Outside, in the blossoming trees, under the cloud-draped moon, a half dozen screech owls had gathered. Undisturbed by our candle, they carried out their spooky concert. From the peninsula came the loud, subterranean grunts of bullfrogs. As we went hand in hand one more time through "our" little house, lightning bugs danced through the night—much bigger and brighter than in our old homeland—by the thousands. From the south quivered lightning, glowing through the cloud banks; all the omens looked like heavy rain, but we were much too tired to worry about that. The surf sang us to sleep.

First Friends

For the next few days we were like Robinson Crusoe, as he explored his island. The first thing Rita did was check out the garden, and before breakfast she came into the house with a big bunch of narcissus: "The old man took a lot of trouble to make his last home beautiful. He laid out a rock garden as well, and he planted rows of iris and narcissus at the edges of all the paths. They're blooming by the hundreds, but you can't see them because of all the weeds and the high grass."

I climbed on the roof and nailed down the loose corrugated roofing sheets, the main reason it had rained into the house over winter. Scattered crazily on the ground lay the old doctor's worthless goods: rags, old newspapers, and even a school notebook. By afternoon we had the little house halfway clean and in order. Our main problem was finding enough screening to cover the windows, because last night the mosquitoes had made us quite uncomfortable; in the meantime we helped ourselves temporarily with muslin.

Then it was finally time that we could go to the peninsula, in bathing suits, in fact, because it was clear that we would either have to wade or swim if we were to get across the ravine where the road was under water. The clay-thick water was lukewarm, but the violent eddy behind Tower Rock had thrown up a barrier of trees and branches in front of the ravine, so that there was almost no current, and we got to the other side peacefully. On the other side of the ravine lay a limestone plateau with an area of perhaps fifteen acres, evidently the last quarters occupied by the quarrymen. Out of an overgrowth of blackberries and man-high weeds stood a smashed gasoline pump and the wrecks of a good dozen trucks. Board platforms, placed a foot high above the ground, were evidently floors for the tents they had put up for the workers; by their size and number I guessed that there had been a minimum of sixty men in the camp. We climbed on a rock, whose top had been blasted off with dynamite—you could see the ends of the boreholes in the stone— and looked over the peninsula as a whole for the first time. In front of us stood Tower Rock, about two hundred fifty feet high and six

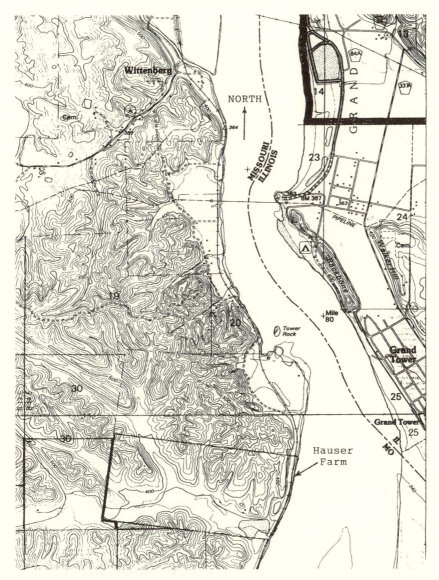

The Mississippi shore showing Wittenberg and the Hauser farm. Map
adapted from the Altenburg, MO-IL (1994), and Neelys Landing, MO-IL
(1993) U.S.G.S. 7.5 Minute Topographical Maps.

hundred feet in diameter at the waterline, its summit overgrown with crippled trees and brush. On our side, the Missouri side, the Mississippi raced by the rock cone at a distance of four hundred feet; the other side of the river was a good two and a half miles away. And since the peninsula from our side, beneath Tower Rock, stuck out into the water a good three-quarters of a mile, it caught the water mass and created a vortex in which the most powerful battleship in the world could not have maintained seaway. This was the surf we had heard. With giddy speed, uprooted trees spun in a radius of several hundred feet, around and around, most of them flat in the water, but some of them upright, with their heavy root-ends below, so that it looked as if a forest was dancing there in waltz time. In the middle of the eddy groundswells erupted, so that trees suddenly shot out of the water like lances hurled by gigantic hands. Apparently not even fish could exist here, because as the trees rolled around and around, sometimes we saw fish fully impaled on their branches. Only in a hurricane had I experienced nature in such wild majesty, and an hour passed before we could turn our eyes from this immense drama.

Sinister black oaks with trunks six feet in diameter and even larger cottonwood trees with huge protruding crowns marked the outline of the peninsula to the east. Broken by it, the stream ran more slowly, but every tree created a wake and was encircled by washed-up undergrowth. At a great distance the houses of Grand Tower, on the Illinois side, looked out over the dike, so low on the water's surface that it looked as if the river would shortly break through; two days later that indeed happened. Downstream from Grand Tower, a tree-grown island, by estimate six miles long, divided the Mississippi; just then a barge tow, fighting hard against the current so that it appeared as if it were standing still, rounded its tip. Mississippi barge tows are completely different from those on the Rhein: The rectangular barges are lashed together into compact larger groupings, or rafts, and the towboat (really a push boat) is fastened to its after end. The whole thing can then be handled like a single ship, but a ship that can be up to thirty-six hundred feet long and that can transport up to twenty-five thousand tons of freight. Almost without exception, modern towboats are diesel vessels with twin screws. Pushing the barge string has an advantage in that the towboat remains in deep water and is maneuverable if the forward end runs aground, as so often happens in slack water.

On the other side of the island, on the Illinois side, we could see a strip of flooded farmland many miles wide. Only the rooftops of the farmhouses could still be seen, and only the tops of telephone poles indicated the roads that were there. It was an unreal sight to see a couple of small paddle steamers maneuvering across this terrain; they were headed for the farms, apparently to rescue the stock. Later, we would learn that in that spot the flood had breached forty miles into the flatlands.

The next night, and for the next few days, escorted by the heaviest of electrical discharges, cloudbursts fell with a force that one could not imagine in Germany. About midnight, we awakened to the squeaking of rats, and in the glow of a flashlight we saw a veritable brook flowing through the house out of the fireplace. The dammed-up water in the chimney had washed out a couple of rat's nests from their homes there. I wasn't in the mood to save lives; it remained for the mother rats to take care of this themselves. To do this they carried their pink babies one by one into the darkness of the room. Twice during those torrential days I had to go to town to shop for groceries, and since driving was out of the question, I walked along the railroad embankment. The tracks traveled right along the river; I remarked to my acquaintances that the stretch was much shorter than the wooded way through the hills. The trek only became difficult about two miles from Wittenberg, where the railbed dipped under water. In fact, the telegraph poles marked the way, but I stumbled and fell often over the invisible rails and crossties; this made no difference, since I was wringing wet anyway. Among the most disagreeable things were the trestles—just as invisible—that crossed a couple of old channels; here, there was no solid ground under the ties, and every little skid meant that I slipped between the ties up to my hips and barked my shins painfully. Half the population of Wittenberg was gathered in Müller's Store; the men drank Coca-Cola at the counter, and the conversation buzzed about the flood. High water was expected in St. Louis the next day, so the crest would reach Wittenberg the day after tomorrow. The fishermen swore over the Mississippi administration, because its dams hindered the holding of the floodwaters further north, and the onrush was always heavier to the south: "The more the government fools around with the old river, the crazier it gets. We used to get a big flood every ten years; nowadays we have to clear out our houses twice a year."

Nonetheless, red-haired Mr. Müller found the time to tell me how especially uncomfortable the disruption of railroad service was for him, because for that reason he couldn't get to his doctor in St. Louis: "I had syphilis in my youth, and now sometimes I have something wrong with my head." This open declaration surprised me a bit, in light of how many women were in the store; but nobody seemed to find anything wrong with it.

Also among the customers were Leo Harris, the old, gray-haired fisherman, from whom we had asked directions in the tavern, and Bill, his black mate. With a friendly wink of his brown eyes, Leo greeted me like an old friend and told me it would be a good idea if I bought some tackle, because the fishing was good at Tower Rock. As I purchased hooks and lines according to his expertise, he suggested that he take me home in his boat; he didn't have to say this to me twice. It was only three paces from the store to the boat; it was tied to a stake really meant for horses. Black Bill began to bail water out of the boat with an empty tomato can—it streamed into the boat not only from above, but also from below—while Leo tossed off the line and pushed us off between the houses, where the current gradually gripped us. This done, he put on his steel-rimmed glasses, held together with string, and began to get the motor ready. It was an old Model A Ford motor, rusted all over, and all its accessories, like the carburetor and distributor, were fastened on with wire. First he wiped the battery posts dry and connected the cables, creating a shower of sparks and a couple of lazy engine rotations. With that, the battery was drained, and Leo went for the hand crank . . . again without result. With a sigh, he hauled out of his pocket a fish knife, which served as a screwdriver, and took apart the carburetor. Meanwhile, we had been driven downstream of the houses; the boat began to swing noticeably and we picked up speed. Black Bill sat, chewing tobacco completely passively at the rudder, and declared, wisely:

"If Leo can't get the motor to run—I don't understand a word about the damned thing."

"The gas line is probably plugged up," said Leo.

Leo Harris and Huc, working on the notorious Model A Ford engine that powered Leo's boat. *Photo courtesy Huc Hauser.*

After a long wire pushed through the tube had brought forth a considerable amount of rust, he cranked the engine once again, and a miracle happened: the machine started. The old flatboat jerked and shot between the trees out into the river. It was built like a flat scow and lay barely hand-high in the water. Leo waved at me to come forward, and there I squatted, straining to watch over the foggy waves for "snags," which threatened at any moment to tear the bottom from under us. Leo nevertheless was an excellent helmsman, and the water that sprayed over me with every wave was bailed back out by Black Bill with his tin can behind me. The current ran at least ten or eleven miles per hour, and the boat as well; so we passed wide of the maelstrom at Tower Rock in barely a quarter of an hour. In a wide arc, Leo steered toward the peninsula and shut off the motor, because we had found still water in the protection of the huge poplars. It was lucky that most of my purchases were canned, for everything else, such as the bread, was soaked through.

The two would not take money, but a drink in the house—well, they had nothing against that. So I brought our guests in to greet Rita, and we had a "housewarming," as such a nice and friendly event is known in America. Here an interesting character trait was revealed, that for the original inhabitants of Irish, French, and Indian blood, and above all for the negroes, is quite typical: Leo and Bill talked grandly about all the work that they had to take on. There was the furniture of a neighbor over in Grand Tower to evacuate, and there was Leo's own house, which was sorely threatened by the flood. But they were far more interested in us than in these urgent matters, and they made no preparations to leave. After we had thoroughly discussed the war situation and after it had long since grown dark, Black Bill took me behind the house to show me where the best earthworms were, and if there weren't any, he said, I should cut up liver and soak the pieces overnight in turpentine; that was the best bait. Leo stretched the new fishing line between the house and the shed—in streaming rain—and showed me how the hooks, a dozen per line, were to be appropriately fastened onto it. There was no shortage of heavy nuts among the machinery scrap on the peninsula, and these were the best sling-weights with which to throw out the line.

The afternoon had thus been a great success and helped us to endure the next day, when the flood nearly reached our house, and I seriously thought about a retreat into the hills behind us, the only escape, since we were cut off to the south as well as to the north.

The next weeks brought nice weather, and with the same unbelievable speed with which it had risen, the river fell—in fact six, even ten feet, in twenty-four hours. On the third day we could walk to the peninsula dry-footed, and the maelstrom at Tower Rock faded into chains of little eddies. The squirrels, which had crept into their nests in hollow trees, came out again, the birds rejoiced, and turtles crept down out of the hills in swarms; and wild doves, as plentiful as swallows in Germany, sat on the railroad telegraph lines. Only now could we appreciate the true extent of our empire. As the trees on the banks emerged as dry land once again, there were trains, especially the two passenger trains at midday; one southward, the other northward, with their white and mulberry-colored dining cars. They symbolized civilization, and, at the same time, they told us the time of day.

When I checked the road and also the Dip on the sixth day, they still appeared to me to be impassable for Perfidio. I was even more astounded, then, when a truck burst through the woods from the south; it carried eight tons of freshly cut railroad ties, and it stopped at our house. A powerfully big man, burnt nut-brown by the sun, climbed from the dented cab as ponderously as a bear, his whole broad face creased with wrinkles. He slowly ambled up to us:

"Hello, neighbors!"

"Hello!"

"Wanted to see who had moved in here; saw the smoke from the hill. Where are you from?"

With this delightful directness he introduced himself to us. Chewing tobacco and picking his teeth with a stem of grass—Ernest Petmiller, who would become a good friend to us and, in a manner of speaking, would determine our fate.[1] He asked for a drink of water; I had cleaned and disinfected the cistern in the meantime, and he tasted it and spit it out: "Good enough," he said, "but the water on our farm is something else." With that, he reached under the driver's seat for a two-gallon wooden keg: "Here, taste this once; it's still cool, I just pumped it fresh."

Rita wanted to get glasses, but our guest shook his head. "Naw. Here's how you do it." He lifted the keg high over his head and let a thick stream of water arc into his mouth. We did it after him, and, in fact, it was the best-tasting we had ever drunk.

A request for a drink of water, even when you have brought some with you, as Ernest had, is a rural custom in America, the best way to get acquainted.

"Do you need anything from town?" he asked. "I can bring it to you. I'll be back this way this afternoon. Or you can ride with me if you want; you won't get through with your car."

This, too, is the best sort of American courtesy. In lonely areas, people help one another with transportation or provisions, knowing well that you are beholden to help the other fellow in return.

Naturally I took the opportunity by the hair, and we drove off. How the man could talk on this unbelievable road, with eight tons of oaken crossties swaying behind, was baffling to me. But he talked nonstop, while I peered anxiously out the rear window of the cab at

1. Henry Scholl remembers Ernest Petmiller well, but remembers his name as *Pepmiller.*

the ties, which looked as if they were getting ready to crush us every time we started downhill.

He lived, so Ernest announced, three miles south of us on the Harnagel Farm. The farm belonged to the Moss Tie Company in St. Louis, a big lumber concern that owned not less than seven and a half million acres of woodland in seventeen states of the Union.[2] He was the foreman of the sawmill and the team of lumbermen who were at that time cutting timber off the Harnagel Farm. Not all the trees— only the hardwoods, for crossties. In two or three months he would be finished with the work. Then would come the winter pause, and they would move to another farm in the Ozark Mountains, nearer to Ernest's home; he was quite happy about that, since he would see more of his wife and children.

I liked Ernest right away. He had the most radiant blue eyes and the friendliest smile in the world. He handled his old truck like a team of horses, with "gittup" when it was a steep uphill, and "whoa" when the brakes didn't hold going downhill. We finally came to the station. "Usually," said Ernest, "the ties would be shipped to the creosote works in St. Louis by truck, but now, since the flood, the Frisco line has to replace so many ties so urgently that they have to be loaded into boxcars."

Perplexed, Ernest came back from the station scratching his head. "My helper hasn't showed up. What do we do now?"

With some doubt in my mind, I suggested that I might perhaps replace his helper.

This appeared vastly welcome to Ernest. Without another word he maneuvered his truck with its rear end toward the open doors of a boxcar.

"Have you ever loaded ties?"

2. This was the T. J. Moss Tie Company, which was bought by the Kerr-McGee Chemical Corporation in the 1950s. In 1995, the corporation sold to The Nature Conservancy approximately eighty thousand acres of Missouri forest lands (mostly in Shannon County), which were then acquired by the Missouri Department of Conservation. Timber cutting remains a major industry in this part of Missouri, though. Other Kerr-McGee holdings were acquired by local wood products industries, and Kerr-McGee still operates a railroad tie treating plant in Springfield.

"Never in my life."

"Main thing is not to let your fingers get smashed doing it. The things are really heavy."

As evidence he held up his own powerful paws, on which indeed whole fingers were badly deformed.

If ever there was a true word spoken about weight, this was it. In their rough-hewn condition, each oak tie weighed between 220 and 250 pounds, and their edges were so rough that splinters tore into your hands with every grasp. By the first twenty ties, when the trip was still uphill, I began to pant, and by the fiftieth I thought my backbone would break. The job wasn't finished with the unloading. The eight tons had to be neatly stacked inside the car. When the pile was breast-high, the next layer had to be loaded by swinging the ties and throwing them upward; Ernest called out "one, two, three," but I didn't let go at the right time and the two-and-a-half hundredweight crashed to the floor at our feet. The summer's heat was still far off, but we were both continuously streaming with sweat, and when we were finally finished, the water keg was completely empty as well.

Since I naturally refused Ernest's money, I was dragged along with him by force to the tavern, where within one hour he drank eleven bottles of beer and I, to my disgrace and shame, only ten. Because of this I almost forgot Rita's provisions, but Ernest thought of them, and that appeased Rita when we landed moderately merry on her doorstep.

Ernest's farm drew me towards it powerfully over the next few days; but naturally more rain showers fell every night, which turned the woods south of us into a morass. No trucks came through here anymore, and, in addition, I had to go to work. At the lower end of the peninsula, near the beach, I picked out the tallest cottonwood tree, a true patriarch, hung with arm-thick vines. I hauled up two straight young maple trees out of the driftwood that lay washed up in an impressive amount, and I made a ladder out of them. Then I dragged boards and slats, also fished out of the driftwood, up the ladder and built a platform, about sixty feet above the ground, in the tree's huge spreading crown. It got a double railing, and from them I stretched my sailcloth hammock, which I had carried with me since my sailing ship days. A wooden box for manuscripts and tobacco completed the furnishings: the nicest workplace I have ever had in my life.

Admittedly, it was an hour every morning before I really got to work. The wildlife around the tree house was too new, too full of new surprises, and too beautiful. When the morning mist rose from the Mississippi, it stood in the air like an endlessly long cloud bank for a long time, exactly confirming the course of the river, so that for many miles, upstream and down, you could see its invisible bends and twists through the woods like a map in the sky. As soon as the first rays of the sun shot like arrows over the steaming water, squirrels, red and gray, scrambled down from their trees to drink, the white herons and wild ducks began their flights, and turtles by the thousands drifted through the water at the banks, their grotesque heads thrust upright out of the wavelets. Under the growing warmth, the woods filled with the most wonderful, indescribably fine scent; it came from the willows, which were in full bloom again eight days after the flood. When the morning breeze rose, the feathery seeds of the cottonwoods floated like snowflakes through the young foliage; this soft snow, which drew a veil between me and the river, kept up the whole day. In the second hour of sun, my tree livened, also. Blue and green lizards used the platform of my tree house as a sundeck; big spotted woodpeckers tapped loudly above me; and between the hanging vines black snakes crept upward—I detested them, even though I knew they weren't poisonous, and I tried at first to drive them away. But there were too many, for my tree was apparently a place of refuge, climbed for purposes of copulation. When it first got really warm, the little hummingbirds whizzed all around me, with a sound like miniature propellers; they untiringly stuck their long beaks into the red vine blossoms, completely undisturbed by my presence. I often observed squirrels sunbathing or having a midday snooze; for this, they sought out really thin branches that swayed them in the wind, laid their heads flat on them, and let all four legs hang off to the sides, like balance poles for holding their equilibrium. For some inexplicable reason they awakened whenever a hawk wheeled overhead. No people ever came to my tree; only fishing boats puttered by occasionally at some distance or shut off their motors and let themselves drift to shore, where their traps hung fastened on long wires. Around noon, I often heard rustling and snuffling; this came from the groundhogs, which had many lodges near me. They are very sharp-eared creatures; with the tiniest movement I made to see them, they bolted upright out of the grass on their hind legs like thickheaded little dwarves. With all these interruptions the work progressed slowly, but I felt as if my whole being had been altered, as if it had, so to speak, sunk roots into the air of this wonderful wilderness, as if for the first time new strength grew in me from this American soil.

On the Horizon: The Farm

At the beginning of May our son came on his summer vacation. He was thirteen at that time, and though I feared he would consider this precaution an insult, I drove to St. Louis to fetch him. The train from New York pulled in, but Huc was not to be seen.[1] When almost all the passengers had streamed by me, something struck me hard in the leg. It was an enormous bundle that ran noisily by me on roller-skate wheels, and on top of this big package was piled a smaller one that somehow seemed to push the larger one forward. At first I didn't think about anything human at all, then I thought it had something to do with a poor cripple; at last the thing raised its head, and it was my son. He had fully prepared himself with a tent and all the trappings for life in a wilderness, and since the equipment was too heavy to carry, he had put it on wheels. He was very proud of his "apparatus."

Naturally, life changed for Huc. For him, much more than for us, the Mississippi landscape was a paradise; above all, though, was the cache of old, wrecked trucks on the peninsula. He was wildly determined to build a supercar, of a construction never before seen, out of the usable parts, and on the first afternoon he came back to the little house streaming with blood because a hand crank on an engine had flown backward against his forehead. Best of all, however, was his .22 caliber rifle. As an "enemy alien" in the United States during the war, I was not allowed to carry weapons, but my son, as a "real" American, could, and so I had bought him the Winchester rifle in Chicago. But since the boy first had to learn to handle weapons, I decided that in this case I was justified in bending the law slightly; the result was that a few rabbits and squirrels fell at the hands of an enemy alien, without the victory of American forces being in any way jeopardized. The addition of meat to our fish diet was greatly appreciated, especially the squirrels, which are a great delicacy, and I much prefer them over chicken. There could not be a more

---

1. Huc Hauser says that his name is pronounced " 'Huck,' as in *huckleberry*."

Huc at age thirteen. *Photo courtesy Huc Hauser.*

comfortable way to hunt: Immediately in front of the house stood a mulberry tree, covered with fruit, which are a favorite food of the squirrels. At the earliest glimmer of dawn, the heavy rustling in the branches told us that there were a good dozen of them at work; Huc had only to stick his rifle out of his bedroom window—he literally shot our midday meal directly from his bed.[2] On the second day we made our way to Ernest's farm. We got around the wooded bog by following the railroad embankment: After about two miles, the road wound its way up the high hill alongside the bank, and as we followed it we found on the ridge of the hill a view as wonderful as the world can offer. From a height of about six hundred feet we looked out over the mile-wide river with the elongated "Treasure Island" in its middle and on both sides the fruitful farmland, dark

2. Hauser says nothing about hunting seasons, hunting areas, or hunting licenses. It may have been legal to hunt on one's own land, but Hauser himself was breaking the law by possessing and firing a weapon, not to mention any hunting laws he may have been breaking.

The Harnagel-Hauser farmhouse, 1998. *Photo by Curt Poulton.*

green with young corn, clear to the blue hills on the horizon, perhaps fifteen miles inland. After another two miles' march through oak and maple forest, the road descended steeply into a valley that stretched down to the Mississippi. Thick undergrowth of bamboo obstructed the view, and to bypass the deeply mudded road we had to clamber along the hillside. At last, however, it stood before us, the farm Ernest had spoken about.

Behind tall maple trees lay before us a broad, white house. It was of the usual wooden construction, but it was built on a foundation of heavy limestone blocks and had the effect of being more sturdy than most houses of its kind. Along the whole front ran a veranda on which an overhanging roof stood on columns, and as we came near, a white dog, evidently ancient, since it only barked once in a hoarse voice, raised itself and then lay down again. On the north side of the house, under a stand of fruit trees, stood an old pump house with a collapsed roof, the usual outdoor toilet, and a stone house that looked quite livable but appeared to be serving as a chicken house. Behind the house, the land climbed to a domed hill, and to the right of the hill leaned a small barn. A boulevard of ancient cedars leading up the hill was so inviting that we followed it involuntarily. To our surprise, we discovered that at the crest of the hill was a cemetery. It had perhaps fifteen graves, half of which were the graves of children; the oldest from 1836, the youngest from the 1890s; thereafter the dead of the farm no longer seem to have been buried on their own land. *Harnagel* and *Dryden* were the most common names; Huc and I were both dismayed to see that some of the large trees surrounding the cemetery had been recently felled, and that in falling they had torn out gravestones that no one had replaced. "Those jerks," muttered Huc, "they're so greedy, they couldn't even leave the dead their shade."[3]

From the hilltop cemetery we looked southward over the farm-house and (judging by the height of its weeds) a large and very fruitful vegetable garden, across all of the farmland, and on to the Mississippi. The view was at once beautiful and tragic. One couldn't have thought of a happier location than this broad lateral valley with its view of the river. But the fields and the meadows stood uncultivated and the wooded hills around it looked like they had been leveled by an artillery barrage. The largest trees had been partly cut down, battered in a manner so wasteful that it would have been considered plainly criminal in Germany. The stumps of these forest giants stood a good three feet high; the lumbermen obviously avoided bending over. The crowns—and what crowns!— lay about randomly everywhere. Masses of wood with which you could have heated whole cities, yard-thick branches that in Germany could have built whole villages, lay completely unused and rotting on the ground. The worst was that in taking down the large trees the woodcutters had let them drop on the younger trees, apparently to break their fall. It was a pity to see forty- and fifty-year-old beeches, oaks, and maples by the thousands bent over with their limbs torn off.

---

3. Since the farm has historically been called the Harnagel farm, one would expect to find Harnagel family members there. However, as of July 1998, there were only three stones visible, none of which bore the Harnagel name, and there is no evidence of the cedar-lined pathway to it. Huc called the lumbermen *diese Schurke* ("these knaves" [or rascals, or villains]). *Schurke* is probably the origin of our slang word *jerk,* which I have used here.

From uphill and far off we heard the sound of a diesel motor rumbling and a circular saw screaming. Slowly, quite taken by the wealth of impressions around us, we wandered in the direction of the sounds. On the left-hand side, in the valley's bottom, stretched a broad creekbed; in Germany we would well have spoken of a riverbed, filled completely with gravel and rubble. The water apparently ran underground; only where the flow had washed out deep holes stood small ponds of crystal-clear water, and since we were sweaty and tired from the oncoming heat, we plunged into one without hesitation. It was ice-cold and wonderfully refreshing; lying in the water, we sensed a light current, so it was not stream- or rainwater that ran there, but real springwater. "Father, everything is more wonderful here than anywhere else in the whole world," said Huc; "I wish we could stay here forever; in any case, let's come here every day."

Three-quarters of a mile farther, a fringe of trees led crosswise through the valley; on the other side lay a large, almost circular clearing, in the middle of which rose a pile of sawdust a good sixty feet high. Towering in front of it were supplies of saw logs, a pyramid of railroad ties, and stacks of boards. A half-dozen trucks and old passenger cars were parked there, and in the center of it all roared the sawmill. A wild-looking woodsman dragged the tree trunks out of the woods with a span of horses; two young men with cant hooks rolled them to the tracks of the sawmill; and an older, German-looking man hauled off the side slabs, or slash, that retained the shape of the logs, to a scrapheap, which burned day and night; at the sawmill stood Ernest Petmiller, who directed the whole operation.

Our appearance broke off the work immediately. Ernest turned off the diesel, and with a hollow sigh the circular saw idled for a while, until its scimitar-shaped teeth finally came to a sudden stop. Then they sat together on logs and empty oil cans, cigarettes were lit, water passed around, and everyone was introduced in proper form. The woodsman, so we learned, was called Crocker and had, as he idiotically acknowledged with proud nods of his head, beaten a man to death a couple of years ago and went to prison for it. The old man was called Baumann. "If one of you gets sick," said Ernest, "Baumann here is the best doctor anywhere around; he knows all about herbs and what's involved in castrating pigs—he's the best in the whole region, far and wide." A redheaded driver, who with wonderful, animal-like elegance strolled over to us, was called Red. "Red, here," said Ernest, "lives in the Ozarks and has fourteen kids at home. He has asked me whether you might know a lot of German fairy tales; he always has to tell his children stories, and his supply has run out." Now, out of the woods and with bearlike slowness, came yet another man, older, a regular bull, with a powerful, hairy chest, naked gorilla arms, a fairly fat belly, and a face so swollen that the wrinkles didn't show: "This is John Hillard," Ernest announced, "our Bee-Father. Yesterday he brought out a bee tree, and the beasts really let him have it; but he brought about seventy pounds of honey home with him. I think he would be happy to give you some for your wife." The bear shook hands with Huc and me with ceremonial slowness; the swollen face suddenly developed a hole in its middle, and a couple of yellowed stumps of teeth peered affably out at us from a bed of chewing tobacco. Where the younger people were concerned, one was called Crocker, the other Baumannsohn. Not much was said about the others.

"So, now you know my bunch," said Ernest in conclusion, "and when you need something, you know where to find us."

That evening, as we spiritedly told Rita about the farm, she suspected immediately what was moving us men most intimately, especially when Huc repeatedly cried, "What a farm! If only it were ours! Then I'd never go back to school. Then I'd never leave here again."

"Are you serious about it?" she asked, when we were alone.

"It's about the prettiest speck of earth I've met on five continents," I answered. "You have to see it. If it were up to me I'd say I was serious, dead serious, but I don't know if the Moss Tie Company would sell it or not."

"How much land does it have?"

"Three hundred twenty-nine acres, Ernest told me. It used to be called the Wine Farm; old Harnagel was supposed to have made eight hundred to a thousand gallons of wine each year—the best wine east of California, the people say. Ernest and his team search in the woods faithfully, looking for the hiding place where the old man is supposed to have buried his best barrels. It's good land; Ernest says it would grow a hundred bushels of corn an acre."

"How is the house inside?"

"We weren't in it; so you'll have to go with us. I reckon it to have at least eight rooms; we could live like princes, if it were ours."

"A royal estate," sighed Rita, "that would be completely beyond our means; forget it, my dear."

In the next few days a peculiar and literally highly explosive tension built up in us all. A heat wave broke over us with an impact that one could not imagine in Europe; under the corrugated metal roof the temperature climbed to 106 degrees by noon. Swimming in the river brought no cooling; in the evenings we squatted around the improvised table, each with our bucket of water, into which we stuck our naked legs. Only very early in the morning could any physical labor be undertaken; from noon on, both the body and mental faculties failed from dizziness. That was the reason why the otherwise cool boy came running home one evening with at least twenty-five pounds of dynamite sticks, fuses, and explosive caps, and I know not what else, and declared, beaming, "On the peninsula, there is a whole barrel of the stuff. It's wet; we'll have to dry it on the stove. When we finally get the farm, we'll want to lay a really big charge in the creek bed—then we'll have a pond."

Without a word, I took the dangerous find out of his hands, and only when it rested on the bottom of the Mississippi did I demand that he show me the rest of the treasure. It was a big metal barrel, stowed in a rock hollow, that contained Charlie Bussen's explosive materials. It had an iron door with a padlock, but the hinges had rusted through, so that Huc could break them off without a problem. The several hundred sticks of dynamite were soaked through, the detonating machine was still connected to its battery, and all of it in the most dangerous condition in which dynamite can be found. I made Huc get the shovel and I buried the barrel. Then I made the boy swear that he would never touch it again, and wrote a postcard to Charlie Bussen to tell him that he'd better come and get his dynamite.

But that was only the beginning of the excitement. We had barely sat down at the supper table, somewhat pale, when Ernest's truck came rattling up to the door: "The war is over," he bellowed. "We just dropped a bomb on Japan, stronger than twenty thousand tons of dynamite."

Thus we learned of Hiroshima. Ernest didn't understand that the atom bomb was something basically new and earthshaking. "Why are you so surprised?" he asked us. "The war is over, even in Europe."

A few evenings later, rockets blazed up inexplicably over Grand Tower from the other shore. Ernest and his truck brought enlightenment: "You all have to come with me! We're having a big fish fry; I've already brought three kegs of beer. Japan has surrendered; there's an armistice!"

With that began one of the most fantastic nights of my life. Rita sat next to Ernest and old John Hillard in the cab, while Huc and I stood behind on the bed, between the clanking beer barrels, holding on carefully and ducking our heads under branches, since Ernest was driving through the dark woods like a maniac; he was evidently already half drunk.

At the farm we were greeted with a big hello. The beer, especially, was welcomed by the whole team. Rita was triumphantly led to the big kitchen, where an oil lamp was burning: "Now we finally have a woman who can fry our fish!"

"Wonderful, but where are your fish?"

"The fish?" The woodsmen scratched their heads in embarrassment: "Yeah—were we supposed to catch them? We thought Ernest brought some with him from Wittenberg."

"You lazy bums!" swore Ernest. "I laid the net out all ready for you. But instead of doing something, you let God be your Samaritan. Now get to it! No problem, Rita; in fifteen minutes we'll have the fish here. You can come and watch if you like."

With our oil lanterns and our net, we went to the river, to the place where the stream opened into the river in a muddy slough. The wooden railroad bridge ran above it; there, the older men and Rita sat down, all of them smoking together like chimneys because the mosquitoes were nearly unbearable. The young men shed their shirts and slipped with the net into the black slime, then they swam out and formed a circle with the net. "Come with me!" cried Ernest to me, "Now the fun begins!"

Splashing in the middle of the net, I discovered that the water was only neck-deep, even when your feet sank into the deep mud. No sooner had I thought I had reached firm ground, when something under me surged upward like a giant spring. It knocked me flat over,

"Rita"—Olga Margarethe
(Laurösch) Hauser.
Photos courtesy Huc Hauser.

so that I had to swim up and out, snorting and gasping. The men on the railroad bridge nearly laughed themselves to death. "Did it tickle you good, Henry?" The net was being jerked all around by heavy impacts. Shouting from excitement, the young people grasped the mesh of the net, and I couldn't believe my eyes as, one after another, they hauled huge carp into the light of the lantern. In masses, the creatures had burrowed into the muck; our footsteps flushed them out. Some of them flew over the top of the net, but most of them were caught in the mesh. This method of netting is illegal; it was also the reason that the men on the bridge stood watch and listened, but the result was terrific: Literally within a quarter of an hour, we had caught about three hundred pounds of carp, among them a few twenty-pounders. I had never before gotten such thumping proof of the natural abundance of this Mississippi land; now it was no longer a mystery to hear how carp are a most serviceable pig food. Rita was unhappy right away; killing and cleaning three hundred pounds of carp is not exactly a triviality. But in America, this was work that the men do; without losing a word, the woodsmen pulled out their pocketknives, and almost as fast as they had been caught, the fish lay there headless, scaled, and gutted. Whatever else was required by way of preparation, compared with these men, Rita was a regular orphan child. These people didn't need a woman to do their cooking; it was just a lot of fun for them to have a lady cook in their midst, whom they could teach all sorts of new things. So Rita learned, surrounded by six giants, how to cut the fillets with a knife at right angles to the bones, which meant considerably easier eating. Then she learned that you never, under any circumstances, fry carp with regular flour, only with ordinary cornmeal. Third, they hauled in an enormous iron pot for her, in which the fat had already been melted over a fire over which you could have roasted an ox. From then on, she had nothing more to do than to toss in the pieces of fish and to distribute the cooked portions among the hungry mouths.

The hungry mouths sat, without exception—absent tables and chairs—on the kitchen floor, each man with a bottle of beer between his legs, a pile of slices of white bread on his lap, and an old newspaper next to him, for the bones. One after another and even-handedly, Rita distributed her offerings; even the dogs, lying between the men, got their share, and so a good hour was passed in wordless lip-smacking enjoyment, until the team energetically insisted that it was time that the cook must eat, too.

By the third keg of beer, old John Hillard hauled out his fiddle. That appeared to be the signal for the high point of the evening; accordingly, we resettled ourselves in the living room of the palace. The rooms of the farmhouse were large and well tailored, but held only beds and a few crates. We divided ourselves up on this elegant furniture; the fiddle came from a homemade case. As a proper "hillbilly," John didn't tuck the fiddle under his chin, but rested it on his belly and fiddled away. Mostly he played "schottisches," and one after another the woodsmen ceremoniously whirled Rita about; then the supply of tunes was exhausted and the host insisted that we had to sing German songs. Rita began with "Am Brunnen vor dem Tore" ("At the spring before the gates") and did her part beautifully.

There was a stormy demand for her to continue, and I was heartily glad to be spared the agony of having to sing. When I couldn't put it off any longer, the right song seemed to be a seaman's shanty, "In Hamburg da bin ich gewesen" ("I have been there in Hamburg"). Old John tried in vain to find the right tones on his violin, but with only distant similarity, and by the second verse the oldest dog raised his head and began to howl in lamentation. My success, or rather the dog's, was accordingly colossal.

About midnight, the beer ran out, but not the festiveness. With a crooked grin, old John magically produced a case of "homebrew" from his crate. The bottles were ordinary beer bottles with crown caps, but from the suddenly serious expressions on the faces around me I knew that something unusual was about to ensue. First, a tin washbasin was freed of its dirt and soap scum in flying haste and placed on the floor. Then, carefully, John, with bottle and opener in hand, approached it, while a helper held a lamp for him. With his face turned away, he applied the opener, and with approximately the force of ten combined champagne bottles, the contents exploded, partly in a great circle, but mostly into the wash pan. After a number of repetitions the pan was sufficiently full that it provided enough foaming brew for a round of empty mustard and cheese jars. We pressed on: It was undoubtedly beer, but of a strength unknown in Europe. Insofar as I could understand John's long recipe, fifteen pounds of sugar to one can of malt and five gallons of water and a few cubes of yeast would produce 12 percent alcohol. This time, in a prophetic foreknowledge of this festive event, he had used no less than twenty-five pounds of sugar; any more would not be good, since what would then result wouldn't be beer, but schnapps. I lean toward the belief that this beer must have been right up against the schnapps boundary, since after some time I observed two phenomena, which, as Rita said, stood in relationship to the beer: The first was my son, slumbering peacefully in a corner between two likewise soundly sleeping woodsmen. The second was a live turkey, which lay at my feet with its legs tied together. I had bought this turkey, Rita claimed, from one or another of those present, for our Sunday roast.[4]

4. Somewhere between folklore and fact is the popular belief that extra sugar raises the alcohol content of fermented beverages. Often, home

It was obviously time to leave. I had nothing against walking off the intoxication, but Ernest insisted he had to drive us; there were far too many copperhead snakes hunting mice along the way, and unlike rattlesnakes, they didn't warn the hiker, and they didn't back away from him, either. "The rattlesnake," said Ernest, "is a gentleman." And since he was one himself, and the night had passed anyway, we must go by truck.

Ernest drove—so it appeared to me at any rate—with unusual slowness and prudence. That might have been caused by the fact that the battery was almost dead, and the headlights gave almost no light, and every few minutes Ernest's eyes fell shut. We already had our little house in shadowy sight when it happened that immediately after crossing the railroad grade, the front wheel of the truck fell into the deep ditch along the side. The impact threw all four of us forward, and the motor stopped; there we stood, with the entire back half of the truck across the rails. Suddenly sobered, Ernest tried to start the motor, but the battery wouldn't turn it. I sent Rita and the boy to the house for flashlights and tried, meanwhile, to assist Ernest with the hand crank. The narrow, muddy ditch, however, offered neither support nor room to move; the beer may also have been a part of the blame, but in any case I couldn't get the machine to start.

"Damn it all," said Ernest, "now it's getting critical; in a half hour the milk train from the Cape comes through; we have until then to get the truck off the tracks."

Rita and Huc came back with two flashlights; Ernest asked if we had any red lamps at home, and when one was brought, he told

---

brewers add sugar, and homemade wine makers add raisins, to make a stronger drink. In fact, if proper brewers' yeasts are used, the rise in alcohol content during fermentation slows and eventually stops the action of the yeast. There may be some increase in strength with the addition of sugar, but the most common result is a bad-tasting brew. The extra sweeteners in no way balance the natural sugars characteristic of the malt or the grape juice used, or the flavors they promise. Hauser and his friends may have confused strong flavor with strong alcohol, but he should have remembered the famous *Reinheitsgebot* (purity law) of German brewers: Beer may be made only from water, malt, yeast, and hops. It seems they all would have been better off sticking with store-bought beer, such as Griesedieck Brothers' or Falstaff from St. Louis.

the boy to put it between the rails a few hundred yards away as a warning signal, with the beam facing toward the locomotive. Then we got out the jack and tried to raise the sunken wheel above the edge of the ditch. Despite placing many rocks under it, the jack sank into the mud again and again. Streaming with sweat from exertion and excitement we went to the river and picked out a suitable timber from the driftwood to use as a lever. As it was now a matter of minutes, we worked at first on the back wheels, in order to force the truck sideways and off the tracks. The timber broke. When we had a new one in place, we could hear from far off the whistling of the locomotive. I will never forget how Rita squatted in the mud and held the lever under the front wheel, while Ernest and I swore and lifted with all our strength on the other end and the boy danced like a monkey, swinging the red light toward the train. We had almost raised the wheel out of the ditch and supported it from underneath when we heard the sharp screech of the brakes; enclosed in a mighty cloud of steam, the huge headlight bore down on us. The engine came to a stop ten yards in front of the truck. The engineer and trainmaster jumped down and looked at the dilemma, shaking their heads.

"My friend," said the engineer to Ernest, quite friendly and without the least excitement, "I can't hold the train for more than five minutes. We'll try to help you, but if your goddamn truck is demolished, it won't be our fault."

Meanwhile, the fireman had brought a heavy cable from his tender; one hooked end was fastened to the front axle of the truck and the other on the engine, and then the machine slowly rolled backwards with a few piston strokes. The front wheel gave a jerk out of the ditch; the back axle jolted over the tracks, and the danger was past. But since the front axle was now pulled up nearly against the rail, the truck now stood splayed on the other side of the tracks, leaning almost to the point of falling over, and it was so near the rails that when the train drove by, it just missed tearing off the outside edge of the bed.

"Ugh," groaned Ernest, "that was a close shave. Can you make us some coffee, Rita? I need some."

Rita could. Thus reinforced, by dawn we were able to build a sort of bridge across the grade ditch, from which the rescued truck could gracelessly be rolled onto the road. When it was light enough to recognize things, I saw between the rails a brown, wriggling bundle. It was the turkey, which I had thrown out in the first excitement over the stranded truck. The train had driven over it without so much as bending a feather; it was only furious.

"You know what," said Ernest, after his second cup of coffee and bacon and eggs, "if we hadn't gotten that damned wheel as far out of the ditch as we did, the engine would have purely ripped the whole front axle off and then it would have simply rammed my truck. That's what those guys do in cases like that. In a word, I'm in your debt. I know you're looking for a farm around here, Henry. The Harnagel farm is a good one, and in a few weeks we'll be through with the tree cutting. If you want it, I'll talk to the agent of the Moss Tie Company. For them, the land as such has no value. In any case, it won't be my fault if you don't get that land cheaply."

Fight with the Squatters

On the next payday the agent of the Moss Tie Company stopped at our house. He was an outspoken businesslike sort, and so I didn't like him very much. After we had started with the usual discussion of the weather and the difficulty of his job—he oversaw a couple of dozen widely spaced timber camps and so he had to put behind him a couple of thousand miles by car—he slowly came to the point: Did I have cash.

"Yes, I have; but it comes to: how much?"

"Now I know that farm; land prices will certainly rise, but the Moss Tie Company has no interest in having land in hand after it has used up its timber value." What would I say to twenty-five hundred dollars; that was dirt cheap.

It was cheaper than I had expected. But since I knew that this offer was only a trial balloon, I did something that inside me broke my heart: I declared that farm to be a bad buy. "The house needs substantial repair, the outbuildings are falling down, and the barn is burnt down completely." It would cost thousands to get the farm running. "All the fences are down, the woods are devastated, and the roads are so damaged by Moss trucks that I will have to build a new road to take produce to market. All the fields and meadows are covered with abandoned piles of tree crowns, and the best fields have been ruined by the slash heaps from the sawmills."

That was all true, but naturally there were counterarguments: "Taken all in all, there are only a couple of roof repairs needed, and the house is sound at its core, with no termites in its timbers in comparison to other farms." Eight-inch trunk diameter is the lower cutting limit of the Moss Company, so basically they only created room for the younger trees; in eight to ten years, the woods would be ripe for cutting again. This wasn't Europe, and I had no idea how fast the regrowth matured.[1] The road was a local one, and the county

---

1. As a point of fact, the Moss Tie Company's deed to the land reserved a right to cut timber there. This was therefore not a good argument in favor of Hauser's purchase of the land.

had the responsibility for repairing it; under these circumstances, however, the Moss Company would do something extra and get the roads in order before it left. The five hundred to a thousand tons of sawdust was not a disadvantage, but a benefit, a benefit for me; it could, by his own estimate, be sold in St. Louis; perhaps for twenty dollars per truckload. In addition, he couldn't promise to sell me the farm, he had many others interested in it.

That evening, Ernest came to ask about the state of affairs: "The price is too high," he said. "Four weeks ago, Red was offered a farm for eighteen hundred dollars, but Red didn't have the money. I'll tell you, Henry, why Moss can afford to sell the farm so cheaply; they bought it in '40 from old Harnagel for seven thousand dollars. That was before the war yet, when lumber prices were low. By my reckoning, Moss pulled at least fifteen thousand dollars in pure profit out of it, double the amount they had in it. If they get, say, two thousand dollars from you, it's free money for them. This business about fast growth is true; in ten to twelve years, the woods will be mature enough for felling again—though nobody knows how high the lumber prices will be then.[2] That business about the road construction is hogwash; for three years, on one side or the other, I have begged Moss or the county to repair the road for me. Nothing happened; they'll do even less for you. Even with all that: If you can come up with the money, you couldn't spend it better; strictly between us, the house alone is worth twice as much." I promised Ernest that I would sleep on the matter, but that was a pretext. What wouldn't let me sleep was the thought that the farm might slip away from me. Rita and I discussed the matter all night. "Why don't you go to St. Louis," she advised, "and deal with the Moss people directly. The agent is only a little guy; he probably doesn't have the authority to close the deal, much less to talk about the price."

Her advice was good. Only seldom in my life was I so filled with earnest and heavy feelings, as on the day I took my leave from the Wittenberg railroad station. It was a matter of all our futures, and I

2. This was a prophetic remark. The two friends had no way of anticipating the enormous construction boom that followed World War II. Also, as it happened, and as stated above, the Moss Company did not convey the timber rights when it deeded the land to Hauser.

was so concerned about the money that burned in my breast pocket that, for greater security, I took the train instead of driving Perfidio. The Moss Company was easy enough to find; it had its own building, a properly imposing skyscraper. In the Land Department, in comparison to the big-city atmosphere elsewhere, it seemed peculiarly rural. Here, visits from farmers were obviously commonplace; my reception was personal and friendly, indeed, nearly hearty, and the older gentleman, who dealt with such transactions, chatted amiably and allowed himself a world of time for me:

"So, you are from Europe," he mused. "How did it come about that the Germans have fallen under Hitler?" He knew hundreds of German farmers, and they were thoroughgoing, orderly, peaceful people, who would never think of falling for such radical nonsense. "How long do you think the war will last?"

After we had discussed the world situation, and he had remarked with great satisfaction that he considered Hitler's "Alpine Fortress" to be pure propaganda, he laid his hand on my shoulder.

"Henry, we don't dwell very much on formality here, so I'll call you Henry, and my name, by the way, is Bill. Henry, speaking just between us, this matter of the farm is a trifle. We own 750,000 acres of land and have 5 million acres of woodlands under contract. We don't have to ponder too long on your 300 acres or so. We would like to give you a chance, but do you in fact have the money?"

"I have two thousand dollars in cash," I said, bravely; "here it is." Then I laid the hot bundle of green notes on the table.

Bill grinned: "You are a proper German. They always do things this way. You want to buy a twenty-five-hundred-dollar farm for two thousand in cash. Well, for all I care—Miss Snyder," with this he turned to his secretary, "would you please ask one of the gentlemen from the Legal Department to come here? And then I'll need the papers on the Harnagel farm." Now I saw the previous history and a map of "my" farm. It was already over 150 years old. That meant that it was already settled during the time of the Spanish reign on the Mississippi, and this again meant that its plat, in contrast to other American farms, was not rectangular, but a polygon that geographically fit the land.[3] In America today this is called a "Spanish

3. In fact, this parcel, today called Survey 2175, is very much like a standard American section (one square mile) in both size and shape, but

survey." In the last century, the farm had belonged alternately to the Dryden and Harnagel families; it had only been the property of Moss for the last five years. Looking through the documents, Bill shook his head.

"The title is clear; you need have no concerns about that. We always have them searched by title search companies before we acquire properties, and here is their report. But now I remember how it was when we bought the farm: Old Harnagel had died the year before; he was a good farmer, even if he drank a little too much of his homemade wine. He had seven big, strong sons, all of them already grown when he died. And what do you know—not one of the boys wanted to take over the farm. That was the reason we got it so cheaply at the time. And do you know what happened to the sons? One of them got so far as to become a schoolteacher, but the other six work partly for us in the impregnation works, and partly as tracklayers for the railroad. That's the way it goes when a family loses its roots in the earth. Think, Henry—you have a boy?" Yes, I'd thought of that—"Let the boy take over the farm when you are no longer there."[4] One is one's own master here, and if he is any good, he can live like a prince; I know the land.

Perhaps it was also like that in Germany a hundred years ago; by my time, though, it was already no longer possible that a total stranger and his business partner would be provided with such a hearty reception and with such good and wise advice. Bill was the incorporation of the best of the old, almost biblical American lifestyle.

The attorney came, and the contract was very soon drafted: Mr. Heinrich Hauser acquires from the Moss Tie Company that parcel of land formerly known as the Harnagel farm, bordered on the east by the Mississippi River, on the north by the Dryden property, on the south by the Hoehne farm, and so on. Containing altogether 329 acres, more or less, for himself and for his heirs to have and to hold

---

it lies at an angle of about 11 degrees west of north. This is because it is riparian to the Mississippi, which forms its eastern border. The river is really the only irregular line surveyed.

4. In a search of the grantee-grantor records on this property in Perry County, I learned that, for reasons unknown, Hauser deeded the parcel to his wife and she deeded it back to him while they were still living there. Upon Hauser's death in 1955, after he and his wife had returned to Germany, Rita deeded it to his son.

as their own property, for the sum of one dollar and other valuable considerations.[5]

This last, odd formulation, according to Bill, was for my own protection. Should I want to dispose of the land at some later time, it would not be desirable to shove the buyer's nose into the matter of how much or how little I had paid for it. I have to believe that the one-dollar clause served the seller's purposes just as much, on grounds of taxation, perhaps.

When the money had been handed over and two copies of the contract had been signed, I was walking on clouds. I couldn't believe that I really owned the farm. The streets of St. Louis were boiling hot; I thought that my dizzy feelings had been stirred up by that, but a stop in a dimly lit, air-conditioned restaurant didn't make the dizziness better. I would have given a lot to have been able to share my happiness with someone, but I didn't know another soul in the big city. The next train to Wittenberg would leave the next morning. I couldn't telegraph Rita, either; there was no one there to get the telegram. I went to the train station, and, to make sure I wouldn't miss the train, rented a room in the station hotel. The window looked directly out into the train shed. There may as well have been absolutely no ventilation. Half-naked, I lay on the wretched bed, sweating and put on edge by the endless coming and going of the trains, the clatter of the masses of people, and the droning of the loudspeakers that announced connections. All of this was the embodiment of that over-hurried, over-mechanized, and too-technical metropolitan world that I wanted to escape. And now I had made my getaway; I had *my* farm, completely outside this world and protected from it. It was an indescribable feeling of exhilarating happiness; it wouldn't let me sleep.

Even though the train didn't leave until 8:55 in the morning, at 8 o'clock I was already standing at the barrier; on the whole return trip I gripped the contract as tightly as I had held the money on the way there. When Wittenberg finally came, and Perfidio stood alongside

---

5. Hauser stated that the tract contained 329 *hectares*. Perhaps he was trying not to confuse his German readers with the difference in German and American land measures. The equivalent of 329 hectares is about 825 acres. The deed verifies that the parcel contained 329 acres.

the track as if arranged, I could then only very softly say, "We have a farm." Shouting wildly with joy, my boy threw himself into my arms, but over his back Rita and I looked at each other with trembling lips: "I knew it," said my wife, "and now our life will be more beautiful, but also more demanding."

And it began with seriousness: Ernest Petmiller had gone away with the majority of his men to locate a new timbering base in the Ozark Mountains; the sawmill stood silent. The only thing left to do was haul away the previously cut wood. But when I went with my son to take the first load of our goods from the little house to the farm, there we found to our great astonishment a fat, red-haired woman with easily a half-dozen children on our property. She turned out to be Mrs. Crocker, the wife of the gorilla-like woodsman who was so proud of having spent time in the penitentiary for beating a man to death. When I asked why she had broken into the house, she answered moderately gruffly: I would have to speak to her husband about that; just now, he was over at the brook, catching a few trout.

I met him there, surrounded by a few more children, his huge paws dripping with blood and fish guts. At first he was quite sheepish: Yes, he had heard that I had bought the farm, but he had let his family into the place, and now it was not possible for him to move out. Where was he supposed to go with his family? All of the houses in Wittenberg had been under water and were therefore not habitable, and there was also no work for him there. As far as I could make out of his mumbled, almost incomprehensible dialect, that was the content of his speech.

I stood there as if I had been hit on the head. How was it possible that a man had simply occupied, with his whole family, a house he didn't own, on my land, on which he didn't even have a job any longer and in which he had never ever had a legal right to reside? In Germany, one would make a big fuss in such a case, or run straight off to the police. With respect to a violent savage like this Crocker, however, that would do little good. In the first place, as far as the police were concerned, there were none available in these surroundings for miles; second, I had always preferred to handle my affairs without needing police and courts to settle them; and finally, such a step would have reduced my esteem in this region, where people are fittingly proud of their independence. As calmly

as possible, I told the woodsman that the farm belonged to us, that
obviously we wanted to move in, that I did not want to make matters
any worse for him and his family than was absolutely necessary. I
therefore gave him a couple of days' time.

Inside, I was deeply disturbed and boiling over with rage, as I
steered Perfidio back toward the little house, preparing to tell Rita
about this. It was a lucky thing that we saw Leo Harris and Black
Bill out on the river fishing; we waved them over and then held
a war conference. To my great surprise, both of these experienced
backwoodsmen took the matter completely seriously.

"I thought that might just happen, Henry," said Leo, "that they
would poke a stick through your spokes." Crocker had planted a little
garden there while he was still working for Moss. That means he had
gambled that the farm would stand empty after the timber cutting
was done, and then claim squatters' rights. In a law that is for the
most part unwritten, but effective practically throughout America,
a man who plants cultivated but abandoned land cannot be driven
off this land without other factors intervening; if he is successful in
bringing in a crop, he earns a property right that is not easily broken.

"But the farm hasn't been without an owner for even a day," I
threw back at him, "and at best Crocker cultivated a couple of square
yards of garden; to make an ownership claim based on that is pretty
near laughable."

Leo nodded his wise old turtlelike head doubtfully:

"You can go to the sheriff, of course, Henry, and get an eviction
order from him. It will be effective in four weeks. If I were in your
place, I'd be careful. As far as I know, Crocker has killed more than
one man. Also, it would cause bad blood in the neighborhood if you
drove Crocker out by force. In your place I'd deal with the man.
If he says he can't find another place to live, naturally there's an
answer that there are empty houses enough in Wittenberg; rent him
a house—that will cost you at most a couple of dollars—and you
will have shown your goodwill and can demand that he clear out."
As little as it pleased me to hear this advice, I could clearly see how
much we would rely on the goodwill of our neighbors in the future.
This time I had Rita, the better diplomat, drive Perfidio to the farm
with an invitation to look at houses for rent in Wittenberg. She went
willingly enough, but when Rita came home from the expedition that

evening she was clearly desperate: "It's impossible to get anywhere with that devil of a woman," she related. "In one house, the stove was unsuitable; the next one, she said, was too wet and rickety; the third was too small; the fourth one, too large for her furniture. I tried to persuade her until I was blue in the face—I bought her cookies and candy bars for her brats, all for nothing: Those people just don't want to do it."

It was the thirty-first of June. The rent on the little house was paid until the first of July; of course, I could have extended the lease, but now I was gripped by anger. I owned a nice house, and it was illegally closed to me: "You know what?" I said to my family. "Tomorrow we're going to take possession, if not of our house, then at least of our farm!"

The next day at noon, as a fully packed Perfidio drove onto our land and ground, all the Crocker kids hurried into the house, apparently to give the alarm. Nevertheless, we drove slowly past the house, and I grinned as I saw, through my rearview mirror, Mr. and Mrs. Crocker peeking around the corner. It pleased me less, I admit, that Mr. Crocker was carrying a shotgun; apparently he believed that we intended to try to take the house by storm. Three-quarters of a mile up the valley, but still on our land, I stopped. A little meadow, well protected from view and not far from a water hole in the creek, offered a nice campsite. We spent the afternoon setting up the tent, putting together the stove, and dragging in firewood. Near evening I purposely shot a couple of squirrels not far from the house, and as our campfire glowed after dark, there was no more doubt on the part of the Crockers of our intent and our visit. The siege had begun, and what I had expected, happened: The squatters sent a parliamentary agent in the form of their eldest son. Right after supper and as we had lit our cigarettes, the young man emerged from the dark woods:

"Hello!" came an unsure voice.

"Hello, Crocker!" I answered with vast friendliness. "Sit down."

The boy squatted on his haunches in the manner these backwoods people had taken over from the negroes and the Indians, and stared at us in embarrassed silence.

"Cigarette, Crocker?"

"Yes, thanks."

As the cigarette was smoked in silence almost to its end, he began: "Don't you like the doctor's house no more?"

"Oh, sure, but the rent agreement is up. We thought then that it would be better to move to our farm."

Long silence.

"Cup of coffee, Crocker?"

"Yes, thanks, if you got one left. Ma told me to say she was sorry that she couldn't find the right house."

"Yes, we're sorry too. It's not comfortable, naturally, when you have to camp in the woods like us and can't get into your own house."

"Pa says he's in the right."

"Could be that your pa thinks he is in the right and the sheriff thinks otherwise."

"How do you know that?"

"Because I asked him myself this morning in Perryville."

"So, you was in Perryville, then?"

"Yes."

"And the sheriff said we didn't have no right to the farm?"

"Yes, he said that."

"How are we supposed to know if that's really true or not?"

I shrugged my shoulders indifferently. "What do I care whether you believe me or not? Just wait, then you'll see. In the meantime, you can ask in Altenburg or Wittenberg what they think about it there."

"And the sheriff knows you're here?"

"Yes, of course; he'll probably come by tomorrow morning to see us."

With remarkable speed the young man was on his feet. "I think I better go home now. Good night."

"Good night, Crocker."

It was a lie, naturally, that I had been to see the sheriff; I had only spoken with a clever old lawyer. It was his advice that I should just worry Crocker a little by mentioning the sheriff. In the meantime he would see to it that his friend, the state trooper, would drive up, just to show himself and to look after the law. Openly confessed, it was fear of a night attack that had brought me to call this bluff. If Crocker thought that the sheriff would come the next day, it was fairly safe to assume that he wouldn't try anything this night. Just the same, I was sorry we didn't have a dog with us in camp, and I slept with the .22 close by my side.

The night passed peacefully.

The next morning we were barely through with breakfast when a white patrol car came bumping up from the south in our valley. The sharp-eyed trooper had seen the smoke from our fire from far off. He stopped his car and ambled up to us slowly, a big, gray-haired man with a friendly face. On his loosely buckled belt dangled his big Colt revolver, and between his teeth was a stem of grass, as was usual here in the country.

"Morning, folks."

Translated word for word, that meant "Morning, people," but in the American manner, not condescending, but in a friendly, "thou and I" sense.

"I hear you're having a little trouble getting into your house."

Yes, we were indeed.

"Now, officially, I don't know anything about it, but it can't hurt if I look in on the squatters and drop a word or two that in the long run they can't stay. These people probably simply don't know the law."

Yes, I thought so, too, and we were ready to settle the matter peacefully, even to pay the Crockers a couple of months' rent, if they would just agree to leave.

We watched tensely as the car drove on to our house and counted the minutes of the visit. It was barely ten, but it worked like magic: The trooper had barely disappeared in the direction of Wittenberg when the younger Crocker came galloping up to our camp.

"Pa and Ma thought things over; they're ready to leave."

"So, you're ready to leave. Now that's smart of you."

"Yes, but there's still a little trouble; before we can leave, we need a truck, and we don't have no money for one."

"Who's supposed to take you?"

"Hellwig, that farmer down in Wittenberg, said he was ready."

"Then let Hellwig know that he can get his money from me, okay?"

He nodded. "Okay. But there's still another little problem. Pa put a lot of work in the garden. There's lettuce there, and beans, and beets, and even a patch of potatoes; he has to get something for that."

"Tell your father we don't want his stuff. He can come and get it himself, when it's ripe."

"Pa don't want that. We need money."

Now I thought it was time to get a little energetic: "I want to tell you something: In the first place, you have absolutely no right to occupy that house. Just the same, I said I was ready to pay you three months' rent and pay for the moving. That is fair enough. Now I'm giving you my last offer: Fifty dollars in cash, if you move by today, with all your stuff, and that afterward you don't come back saying you want this or that. If you don't want to accept that, then in a couple of weeks the sheriff will come and drag you off and you won't get a cent."

"So, that's it: Fifty dollars in cash if we leave today, or in four weeks, the sheriff and then nothing?"

"Exactly. You understood it."

The boy ran back to the house like a startled rabbit, and shortly thereafter we saw on the distant veranda vigorous and, for us, merciful activity.

That afternoon, Hellwig's ancient Ford truck rolled up, and they first had to fix a broken spring and damaged driveshaft. When those were remedied, one of the Crocker kids came running up to us again: "They're having trouble catching the chickens; will you come and help?" We did that, and the squatters—in anticipation of their money—were overflowing with apologies for the damned chickens. But since the poor birds had no chicken coop, they had gotten used to establishing their nighttime quarters in the trees, and since you couldn't catch a wild chicken, you had to wait until they went to sleep. When we finally plucked them off their roosts, with the help of lanterns, and stuffed them into sacks, the family father was missing. The concern of his people was great, namely because he had gone off to fetch his fishing lines, along with his gin bottle. I went off on the way together with Farmer Hellwig, who swore and cursed that he would never again transport such a band, and that he had missed his supper. After a long search, we found the gorilla slumbering peacefully on the railroad tracks. He had rested his neck on the rail; next to him lay the empty gin bottle. Immediately we heard the sound of the signal from below, announcing the approach of the evening train. Fat Hellwig shook his head and stared at the helpless man.

"Somebody else should be here to say that dear God doesn't watch over drunks. If we hadn't come now, in three minutes, this bonehead would be lying there crushed like an egg. No loss to the world if it happened. On the contrary, the hangman would probably be spared a bit of work, because in the end this guy will go to the electric chair."

"But we're not him," I said. "Come on Hellwig, I'll take the shoulders and you take the feet, now, then."

It was the heaviest corpse I ever helped to carry, and when we stowed the drunkard between the furniture in the back of the truck, the red-haired woman next to us wasted little thanks on us. As she was handed the dollar bills in the glare of the headlights, she growled that there must be something left in the bottle, and that if ever in her life she needed a drop for her weak heart this was the moment for it. Never in my life have I said to a person such a hearty "Have a good trip," and as the truck drove off, we stood there listening for when it made it up the steep hill. Only when we heard the motor noise echoing through the woods, we said as with one mouth, "God be thanked."

It was too late, and we were too wrung out from all the stress to do very much that evening. While Rita and the boy threw open several windows, to let out the odor of unwashed bodies, I drove Perfidio in front of the house, and set up the cots on the veranda. In the meantime, we couldn't sleep long anyway. Again and again I saw a new match light a new cigarette.

"Why can't you sleep?"

"But I am sleeping. In any case, I'm dreaming."

"What are you dreaming about, then?"

"I'm dreaming that we are on our farm."

"That's no dream; it's the truth."

"I know, but it's too nice to be true; let me dream."

"Tomorrow," I murmured, "tomorrow will be a big day."

We Take Possession

I'm not sure which had first priority in the early days—the unending labor, or the unending sense of happiness. Huc and I wandered through the woods, exploring our land and getting to know its boundaries; we discovered many new heights from which we could look out over the river, and over and over we shook hands, saying, "We couldn't have found a prettier place," or, "As long as we live, we could never grow tired of looking at all the wonderful things that belong to us here." When on the way home we met Rita collecting berries, or under a wild cherry tree whose fruit had shaken off, there was always something new to report—of hazelnut groves nearly ready for picking, of brooks in whose beds walnuts lay by the hundredweight, of a vegetable garden hidden in the woods not far from the house, of blueberries in enormous quantities, of beech mushrooms, of pines, and of a dark ravine in which chanterelles and goatsbeards proliferated.[1] We were shaken by the wealth of our land, the ease with which one could live off the "fat of the land," and the impossibility of the idea that one could harvest so little of the blessings of nature. When we headed for home, three in one, carrying Rita's basket and with all our pockets filled with samples of the things we had discovered, then at some bend or another our house appeared, and we sat down at the edge of the path, just to wonder at it.

"Isn't it splendid to look at, our house?"

"Have you ever seen such heavy timbers, such broad boards, and such a massive foundation as in our house?"[2]

"No, nowhere. And how friendly it looks with its shiny window-panes, and how invitingly the smoke curls from the chimney; it

1. Hauser used the word *Ziegenbärte*, "goatsbeards," a German common name for edible coral mushrooms. His *Birkenpilzen*, or "beech mushrooms," are probably edible boletes.

2. Dorthy France states that the original structure of the house was of two single-story log cabins with a dogtrot passageway between them. They had later been enclosed in board siding and the second story added, and the dogtrot was closed by doors at the front and back. Also, an indoor kitchen was added at the rear.

knows that a proper family has saved it again, and it thanks us for
our care."

As good luck would have it, we also had things that came to us
via the most important and most colorful book in America: the mail-
order catalog from Sears and Roebuck. In wartime it wasn't as thick
as it was in time of peace, and even though we had to be absolutely
frugal, the most important implements arrived little by little. Each
new order was first unpacked on the veranda, a little Christmas every
time, and was set up and tried out.

A single-barreled shotgun for more efficient hunting; it would pay
for itself in a few weeks.

Wonderful axes, mattocks, and brush hooks, their hickory handles logically painted red, so they couldn't be so easily lost in the
woods; now we could finally go after the bulk of the tree crowns, the
undergrowth, the overgrown gardens, and the rutted roads.

A two-handed saw made of bright steel, with such a wonderful
ring that it could be used as a musical instrument.

A double-action pump with a sixty-foot steel casing, so that we
could put our water supply on a completely new basis, with a
pressure feed into the kitchen.

Washtubs for Rita, a canning apparatus, a closure device for tin
cans, and a supply of cans, which were at that time difficult to get.

Rolls of copper screen, to make our house mosquito tight, and
heavy rolls of roofing paper for all the new roofs we planned.

Big steel cans that felt like lead, with white paint for the house,
pastel colors for its rooms, brushes with them, and canisters of
turpentine, along with sacks of plaster for patching the walls.

All of this was a fortune, precisely because it all meant an immeasurable amount of work, and work and happiness were one.

Only one thing was missing to make us all completely happy,
which Huc nailed down on the very first day as an acute requirement:
"There has to be a dog here." There was no shortage of every sort
of mongrel in the village—an offspring of that sort we could have
gotten easily and free—but this time our pride reached a little higher:
It had to be a real hunting dog, one that could chase squirrels to the
trees and then bark them out, one that was smart enough to go on
raccoon hunts with Huc at night. Only one breed in the area could
perform like that—the true bloodhound. There are many sorts of

bloodhounds, and not all of them look so dangerous as the ones represented in *Uncle Tom's Cabin*. The bluetick hound, for example, looks almost exactly the same as the German *Vorstehhund*,[3] only its clever head is a little longer and narrower, and the color of its coat is dark gray with silver-tipped hairs.

One of the truck drivers, who continually hauled away the mountains of ties for the Moss Company, had a bloodhound kennel at home. Permission to hunt on our land had made him well disposed toward us, and after Huc and I had helped him load ties a couple of times, he promised to bring us a pair of eight-week-old puppies, females, because Rita wanted to start her own bloodhound kennel.

On a burning hot afternoon the red truck came rattling up to our place after a two-hundred-mile trip, and—oh, what a pitiful sight met our eyes!—the driver had let the pups crawl around without restraint between his feet and the pedals; tossed about, beaten up, and three-fourths dead from thirst, there lay the little bundles of unhappiness, whimpering softly.

"Doesn't matter," said the fellow, robust to the point of brutality, "they'll recover fine. Bloodhounds can take anything, even snakebite." Rita almost cried, as she carried the little things into the shade. With their thick, shivering heads sniffing around them carefully, they cuddled shyly in these strange hands and didn't want to take water for a long time. Only after they were left completely alone for a time did they totter around on shaky legs, as if they had to convince themselves slowly that this was really firm ground, and after many distrustful examinations of the strange surroundings, they took their first nourishment around evening. The stones fell from our hearts, as we saw the red tongues lapping, and then, their little bellies tautly filled, how they became curious and very slowly tame. They whimpered for their mother for three nights, because it is the nature of the bloodhound that he is perhaps the most loving creature God has ever made. From the beginning they chose the space under our veranda as their safe haven—the bloodhound isn't a house dog by nature—so we therefore placed their box there, which they grasped immediately was home, after they had flatly refused

3. A pointer or setter.

the living room. A few nights later fell one of the cloudbursts that is so typical of the Mississippi, and we awakened to the most miserable cries that had ever struck our ears. As we ran outside with flashlights, there we saw a brown flood that covered the whole front yard, and in the middle of it, driven by the stream, the dog crate with the yammering puppies in it, like Moses on the Nile. The rescue operation was over in a flash, naturally, and from that blink of an eye on, we felt as if the animals were really ours. The last barricade of shyness had been broken down, and they quite obviously viewed us as their lifesavers and showed endless gratitude. From then on, they slept on, not under, the veranda, and since Huc had saved "his" dog first, "Kitty" slept in front of Huc's window, while "Holla," whom I had "saved," slept under mine and Rita's. Bloodhounds are that clever.

Work? It was simply endless, and there was work to do first in order to make work possible: namely, the clearing away of five years of devastation. Because of the heat, we did the heaviest work in the earliest hours of the morning. Huc, who had become a "right good" driver, started Perfidio and drove it either to Cemetery Hill behind the house or to the field next to the house.[4] Meanwhile, I was already out with cant hook, saw, and axe, to cut fallen trees free from their roots, to trim their branches, and in general to make them transportable. Huc maneuvered the car backwards toward the respective tree and we attached the tow chain; the boy drove forward, and nearly always quite sizable trunks were moved and vigorously dragged to an eroded ravine nearby, where we used them for fill. The old Packard, with its high wheels, its low gear ratio, and its hundred horsepower, worked like a truck; it couldn't be stopped. When we got to the ravine we would unchain the tree and roll it into the ditch with the cant hook; the more trees we filled in, the better, since they hindered further loss of soil. In this way we were able in only a few

4. It is difficult to see how the first of these might be possible. Cemetery Hill, simply put, is impossibly steep to climb in a car, even in "Perfidio," the miraculous 1929 Packard. For that matter, it is difficult to imagine pallbearers carrying coffins up that hill, although there was perhaps a way in from the west, through the woods.

weeks to clean away the masses of wood left by the sawyers from
Cemetery Hill, the garden, and the nearby fields. Using this material
other than for fill was unthinkable. Firewood, that you would first
have to cut, was available in the area for ten dollars a truckload—five
tons, that is. All around us in its own right, lay heaps of stove-ready
slash from the sawmill that would last for years. One cannot think
of a greater contrast with Germany than this.

After breakfast, a hearty breakfast with fried eggs, squirrels' livers,
fried potatoes, and many cups of strong coffee, the dew was dried off
enough that we could begin mowing. I don't in any way overstate
when I say that the weeds around our house were a good six feet high,
and often so tough that they couldn't be cut with an ordinary scythe.
A special scythe with a shorter, heavier blade and an especially strong
handle was needed, and it was heartbreaking how often we had to
stop to sharpen it, because rusty wire from fallen fences ruined the
blade over and over. We also usually found more tree trunks and
stumps while mowing, because they were hidden from view; we
had to haul them away then. If the wood was too rotten, so that it
shattered during hauling, we had to set fire to it in place. We could
only do that after heavy rains, though, because of the danger of
wildfires, and even then we had to be careful, because the old trunks
would glow as embers for days, and any strong wind could ignite
a forest fire. By nine o'clock in the morning Huc and I, clad only in
denim pants, were scratched up, covered with burrs, encrusted with
dirt and ashes, painted with bark and sawdust, and therefore were
over and over again partially washed clean by streams of sweat.
But the feeling of triumph of having created a halfway orderly
yard out of the wilderness around the house cannot be described
in words.

At about eleven-thirty, a half hour before lunchtime, we stopped
and threw ourselves, out of breath, into a marvelous water hole in the
creek bed. Towels, scrub brushes, and soap lay ready for this ritual
in a hollow tree. Then we heard the gong ring from the house, an old
plowshare hung from a tree, and with true threshermen's hunger we
sat ourselves down at the round table Rita had made out of the crate
from a circular saw. Fried squirrel was the most common and almost
the best-loved dish; we enjoyed rabbit only when necessary, when
the hunt was poor; Rita's fishing lines provided us richly for at least

two or three days a week; we only had meat from the store when the work was so demanding that there was no time to hunt.

After lunch, I let the boy go hunting; it was his summer vacation after all, and after a hard morning he had certainly earned the pleasure of the hunt. We only set as limits the borders of our own land, and a series of three shots were to tell us that something had happened to him. Since the hills all around echoed the sound so wonderfully, we almost always knew where he was, and, God be praised, the alarm signal never had to be used.

After a couple of cigarettes in the shade of the veranda, I went to the garden, and it wasn't always easy to get my tired bones going again properly in the afternoon heat. In point of fact, the vegetable garden was much too big for three people, and even though we started cultivating it far too late in that first year, we still could have fed ten families. To make garden land that is totally covered with weeds productive again is no small matter; even out of freshly dug earth, the weeds shot up in such numbers that we couldn't cultivate with a hoe and cut back once a week or every two weeks as is usual in Germany, but rather had to do it every second day. I had not had to worry about fertilizer, and despite my not having used any, every tomato plant bore between fifty and a hundred pounds of fruit— tomatoes weighing two pounds were no rarity. Pickling cucumbers had to be harvested every two days and canned, because within a week they would have grown to the size of salad cucumbers, and any later they were fully useless, lying on the ground brown and yellow and the size of melons.

As soon as the heat let up a little around five o'clock, I climbed up a homemade ladder to our corrugated steel roof. For a half-dozen years, nobody had done anything for this roof; twisters had loosened the sheets, and it was easy to foresee that they would soon rust completely through. The work up here was just as boring as it was hard. The rust was scratched off with a wire brush, the edges of the panels renailed, and then it was painted with red lead paint.[5] Since

5. Lead paint was effectively banned from residential usage in 1978. Red lead paint was, however, very effective as an antirust and all-purpose primer and was used everywhere as both primer and color coat for farm structures and other buildings.

the ladder didn't reach the second floor, I had thrown a long rope over the roof, one end anchored to the ground and the other end tied in a sling that held me in place. Rita didn't love at all to watch me at this work, but it was absolutely necessary, and, from my old seaman's time, I had much more practice at hanging from a swinging line than the boy.

It was marvelous when, from my rooftop, I could see the boy coming home from far off, waving his game back at me, and pantomiming where and how he had shot them. I would clamber down quickly, and then, in the evening, Rita would come with us when we went for the second time to bathe in the creek. It was inexpressibly beautiful, when all around our water hole wild doves cooed in the trees (I had placed a ban on shooting them anywhere around the house) and how we then, all pores prickling with coolness, strolled into the evening's glow, with the puppies playing at our feet, back to our house.

Yet neither sundown nor the evening meal meant an end to the workday. In those weeks, Rita, along with cooking, washing, canning, and making furniture, had to clear away five years of abuse of the house, from upstairs down to the cellar, and she saw to it that every evening big boxes and trash cans full of rubbish awaited us in the front yard: empty whiskey bottles, gin bottles, beer bottles, rusty tin cans, rags, junk, and who knew what. All that had to be hauled to the ravine or burned, accordingly, and it was honestly grotesque, how out of every corner new trash saw the light of day. I believe it would be no overstatement that in those first weeks we carried away five truckloads of trash and junk from the house and yard. That lasted until it was quite dark, and we had to light the "Aladdin" in the living room. The Aladdin is a cross between an oil lamp and a gaslight; the last attempt of petroleum, so to speak, to win the battle over electricity. These lamps give a very bright, white light, but require a careful warming of their incandescent mantles, and give off so much heat that one doesn't like to use them in summer.[6] Mostly we were too tired to talk much: Huc leafed through the Sears and Roebuck

6. Aladdin lamps are still made, in Nashville, Tennessee, and are in common use today. They are finicky and fragile, and, as Hauser stated, must be warmed up and adjusted carefully, or they will smoke badly and

catalog, Rita sewed, and I stared into the darkness, where big moths flapped against the screens over our windows. The whispering of the river and the dull chorus of the bullfrogs penetrated in to us. Occasionally a hoarse bark would come from a faraway hill, and our dogs growled outside on the veranda. Then we knew that there would be rain in the morning, because the barking of timber wolves at night announced bad weather.[7]

So, fall came slowly, and the pain of parting. Two years before, when we moved to Chicago, I had placed Huc in a good boarding school in the state of New York. He had made good friends there, and he was a good learner. It would have been irresponsible of me to keep him on the farm. The nearest high school was in Perryville, almost thirty miles away from Wittenberg. Of course there was a bus for the Wittenberg and Altenburg children, but the boy would have had to walk twelve miles to and from the farm every day, and I knew that our road was practically impassable in winter.[8] So there had to be the separation. We didn't speak of it. Thanks to his happy and carefree nature, our son didn't think about it, either, until almost the last day.

On that day, however, a serious disaster almost occurred.

Since his departure, thank goodness, we hadn't seen much of the squatter Crocker, although we knew he was still nearby. The Moss Tie Company had given him the job of dragging tree trunks out of the last cut-over woods. For this purpose he had been given charge of a team of horses that were kept in a stall in a small shed at the sawmill. Crocker was supposed to take care of them. Three or four times it happened that we heard the complaining neighing of these horses.

---

coat their incandescent "mantles" and chimneys with thick soot. Properly adjusted, they give a light equivalent to that of a sixty-watt electric bulb. The reader may be pleased to know that the farmhouse has an Aladdin lamp once more: The farmhouse still has no electricity, so I gave Dorthy France an Aladdin for lighting and warming the house.

7. It is likely that Hauser was misidentifying the canines he heard, because timber wolves had been essentially exterminated from the eastern United States by 1900. Coyotes, on the other hand, remain in the state and have been known to bark and howl before storms.

8. The actual distance from the farm to Wittenberg is about three miles. Hauser's comment about rough going in winter was quite true, however.

The gorilla was too lazy or too drunk to leave the village to attend to them. The helplessly tethered animals were neither fed nor watered. Naturally, Huc had taken over this responsibility, but every time he came home he had tears in his eyes. He could no longer bear to see how Crocker left the animals unattended after driving them to the point of collapse. In the night before that last day it had rained hard. We couldn't do much work outside, so when the boy expressed a wish to take one more long last walk "all the way around our farm," I was naturally in agreement. Right after breakfast he took off. Then it happened that we heard fearful, roaring swearing that sent a chill down our spines. I ripped the shotgun off the wall and ran in the direction of the noise, Rita behind me. I myself was swearing now because I hadn't thought to take the car. Suspecting the worst, I ran nearly as fast as I have ever run in my life. And as I finally rounded the tip of the woods, I knew what had happened: Pressed up against the wall of the shed stood my son, deathly white. He had no weapon with him. In front of him, one great paw clutching Huc by the chest and the other huge hand balled into a fist raised to strike, stood the woodsman, furiously stamping and roaring. I didn't wait one second before aiming the shotgun: "Crocker, one punch and you're a dead man!"

For a moment, he looked as if he wanted to jump on me; then his prison experience won the upper hand. The fist slowly lowered, and the huge oxen head drooped between the massive shoulders. Like a retreating thunderstorm he disappeared into the woods, growling and swearing. As Rita kneeled before Huc, the trembling boy was unable to speak. I knew what had happened: The brave little fellow had confronted Crocker about the horses and had come within a hair of paying for it with his life.

And so it happened that we went to the train station with thanks in our hearts for the protecting hand that watches over children. And as our son threw himself, sobbing, into our arms, the ache of separation was drowned out by the very great feeling of good fortune that he had been safeguarded.

Winter Cares

As the train disappeared around the curve, as we were sitting in Perfidio once again and Rita was wiping away her tears, I took her hand:

"Now we're all alone, darling, in this wild place,[1] and fall is coming. We're going to have to pull ourselves together with strength and support one another. We know this from long before: Madness prowls behind such loneliness."

She shook her head. "No, not this time. We bought this farm for a very distinct purpose. It was to help us to provide help for the people in Europe. That's a goal and a duty so great that we surely won't have to think of ourselves that much."

It was hard for us nevertheless, to return to the house, to the whimpering dogs that looked everywhere for the boy, to his bed, which still bore the imprint of his body, and to the .22 rifle on the wall that I had been ordered to keep well oiled, "And don't forget, Father—at sixty paces it shoots a little bit to the left." We hurriedly washed the dishes with the barely touched plates; the sign that we were supposed to have been together at that last meal . . .

And then began the real reflections on farm life. For a long time, I had known how impossible it is to make clear to the people in Germany what being alone really is like. The friends and relatives we wrote to about the road leading to our farm automatically assumed that vehicles and people must be moving on such a roadway constantly. They could not imagine that through long winter months absolutely no cars and no people came by; that if in November or December a tramp wandering south came to our place, he would seem like a ghost to us. In places like this, as it was in prehistoric times when there were no humans, nature is so overpowering that

---

1. Hauser used the word *Wildnis,* "wilderness," which a German might well have used to describe the area upon seeing it for the first time. Later, Hauser explains how impossible it would be for Europeans to envision such solitude and isolation. Nowhere in western Europe are humans so separated from one another as they would be here on this farm and in this place.

83

an individual begins to doubt his right to be there. This was how it seemed when the fall flood came and the river creatures fled for the hills; when hundreds and yet again hundreds of snakes crept over the farmland; as almost step-by-step the roaming turtles pulled in their heads, hissing, and the squirrels sprang from treetop to treetop in such great numbers that the branches were shaken crazily. Then, indeed, you realized to whom this land really belonged. That feeling was even more overpowering when the wild geese began to migrate; they flew over through all of November and half of December. From the earliest light of dawn until late at night, their wild, lamenting cries came from high in the sky—the most wistful cry in nature. And looking up we saw row upon row of them migrating southward, like snakes winding through the sky. The sandbanks of the island in the river were speckled with their dark bodies, and when hunters hit them and came to collect the downed birds out of the stream, their boats exploded out of hidden coves, with howling outboard motors, foaming wakes, and the bangs of shotguns. More than three million wild geese migrated down the Mississippi Flyway that year, and their passage went relatively unmolested because most of the hunters were still in the various theaters of war.

Paradoxical as it might seem, our feeling of being cut off was sharpened as the first letters from Europe began to trickle in slowly. At first Danish friends wrote, then the French, then the Hollanders, and finally at last the Germans. The mail reflected the end of military campaigns, the end of censorship, and the states of mind the war had created. It was clear to us what a monstrous and horrible experience we had been detached from through all those years. People had suffered such innermost wounds—and many from the recently freed countries wrote candidly that it had taken strength of will for them to express themselves to Germans, even though they knew that we had been in America throughout the war and had taken no part in the crimes of the regime. Then came the newspapers, with their horrible reports of starvation, unforgettably frightful pictures from the concentration camps, and the horrible damnation of everything German, this coming even from the clergy. We were often so discouraged we thought it might have been better to have remained in Europe and to have shared the fates of those who lay under the bomb ruins.

There was only one cure for this depression: work. And this work could have only the purpose of doing as much as possible to diminish the hardship in Europe. The most important job, therefore, was to get the farm into production. Since fats and oils were the most needed foods, I began by organizing a hog breeding operation. As an enclosure I chose a ravine in the woods that would easily become a watering place and a wallow, because so much water flowed there after every rain. As far as possible I used the standing young trees as fenceposts, but there were still hundreds of cut posts required, and I aristocratically cut them from young cedars. A cedar post can stand in the ground for up to half a century without rotting. Left over from the tumbledown and overgrown fence that old Harnagel had once built there were hundreds of feet of still useful heavy-gauge hog wire. This was by no means enough, though, and since there was no money, Rita and I built a full quarter of a mile of woven fencing, principally out of young maple trunks. We laid the woven fence around the steepest sides of the ravine, as we thought that the swine wouldn't fight against it so hard on a steep slope as they would on flat ground. On the flatter ground, we built the pig houses with elementary simplicity, thanks to a clever idea of Rita's. At that time, the Frisco line was changing out crossties, and we had observed that the track workers simply shoved the old ties off the bank and into the Mississippi. We spoke to the foreman, and he said he was willing to stack the ties on the other side, where we could come and haul them off. This would cost his men a little more labor, but we paid for it at five cents a tie. In just a few days, I was able to drive away with several hundred of them, dragging them behind Perfidio a dozen at a time. Out of them I built our hog houses in Lincoln Log fashion. Thanks to the weight of the ties, I didn't even need to nail the walls together. The sawmill furnished an abundance of side slash for the roofs. We had no straw for bedding, but I saw no reason why the swine couldn't sleep on sawdust, and since we had whole mountains of it, the problem was solved very quickly.

Late fall is not an advantageous time to buy pigs. Meanwhile, after many inquiries, we heard in the Altenburg bank about a widow who wanted to move to the city, and we bought four open-range pigs from her. The old woman's farm was in complete neglect, and the pigs ran wild over all the land. While she baited the pigs with the food

bucket, we and several neighbors lurked in the bushes with ropes.
There was an exhilarating fight until we had two of the animals in a
stake partition and two more in Perfidio's backseat, where, although
they stank more than usual because of fear, they resembled indignant
old matrons. Thank God I had made the gate to the pigpen broad
enough so that I could drive in the car and accomplish the unloading
without catastrophe. Twenty-four hours later, each and every one of
the untamed Poland-China pigs had broken out. We found the place
they had escaped and repaired the damage, but nevertheless the
pigs weren't there. Even though I searched the grounds for miles
with the dogs, I saw nothing of the animals and heard not a sound,
all day. Then we thought that maybe they had run the fifteen miles
back home, or had swum over the river, so that the hundred dollars
we had paid was completely lost. Then, on the third night, I heard
grunting directly next to the house. We stumbled outside to chain
up the dogs first, and we listened with indulgent love to the sounds
of our long-lost children, who appeared to be digging up potatoes
in the garden. Well, at least they were here!

The next few days were spent trying to catch them. Troughs set
out in the pen with the most tempting baits failed; no, as freeborn
Americans they would hear absolutely nothing of incarceration.
Feed buckets set out at a respectful distance from the house they
accepted well, but as soon as you tried to drive them toward the
house, they scattered squealing away. Again, it was Rita who found
the cleverest solution: "The fish traps," she shouted, "run and fetch
the fish traps!" I hadn't known until now that one caught pigs in
fish traps, but, dubiously, I did what I was told to do. We owned
four such traps, and we suspended each one from a tree at the center
of the narrow end, in such a manner that its wider opening swung
about a yard above the ground. The nets hung from long suspension
lines that led back to the house and were loosely fastened to the
veranda posts there. Then food buckets were set under the nets, and
the traps were ready. When everything appeared peaceful and the
dogs were tied up, the expectant swines' grunts came nearer and
nearer. Carefully checking to see if things were safe, a pointed pig's
nose peeked out from behind the farthest tree, and then there was no
holding back: With joyful squealing, the slop bucket was overturned
and a powerful chorus of smacking arose. I let the line go, and the

net rushed downward with two pigs in the trap. How the dogs got loose at this moment, I don't know to this day. In any case, in the next second, two Poland-China pigs, two bloodhounds, and both of us were all tied in a wild knot. We finally succeeded in dragging the wriggling mass, with the dogs hanging on their ears, into the pen. In the afternoon we caught the second pair in the same way. As we went to sleep that night we thought we had won. But it was a victory at too high a cost.

The next morning they had broken out again. Rita and I raised and reinforced the fence until by our reckoning neither a mouse nor an elephant could have gotten through. The traps were set up again and the wild pigs—that's what they were, really—were caught again. And again they broke out. Rita swears to this day that she saw with her own eyes how one of these refined swine had taken a board in its mouth and leaned it against the fence to use as a ladder. I take this as a slander, though, because to this day Rita can't stand pigs. Only after the third capture did they become tame. Every evening I stood in their midst with a big sack of walnuts, which lay around us in the woods by the ton, and as fast as I could crack them with a big wooden post, they snatched them away. We became the best friends in the world. I personally love pigs, and in view of their similarity to humans I find it quite unfortunate that we have chosen them as our favorite food. Meanwhile, my friendship was nothing in comparison with the hog-love of our dogs. Kitty and Holla not only got used to being with the pigs, but even slept on them. Nothing was more comical than the view on a frosty morning as the dogs shied away from touching the cold ground and with great stateliness enthroned themselves on the backs of their favorite pigs. After only four weeks we could leave the gate open, and with one evening meal of cooked potatoes we had absolutely no more feeding concerns. The swine foraged freely over the yet uncultivated farm and ate themselves stout and fat on acorns and nuts, and when night fell, we had only to say to Kitty and Holla, "bring the pigs," and prim and properly they dragged their darlings back home by the ears.

Our first animals included—unfortunately—also the Moss Tie Company's team of horses. The work of dragging out the tree trunks was finished. Crocker, the woodsman, had been fired and had moved away with his family. The horses, however, had simply been left standing in their stalls in the field barn without a penny's concern over what would happen to them. My first thought, naturally, was simply to turn the poor animals loose, so that they could at least find their own food and water. At the end of November, all the grass was brown and dry, and we ourselves had no fodder. I wrote the Moss people a postcard, suggesting that they might come and haul off their horses, but a pair of worn-out, almost worthless old nags were apparently of no concern for such a grand firm, and nothing happened. I had bought a sack of corn so they wouldn't starve. We gave them a couple of ears every evening. Corn was expensive, though, and the unfortunate creatures were so emaciated that they rushed at us whenever we showed ourselves. Rita got somewhat angry about it all, and as I couldn't watch their hardship any longer, I wrote to the Society for the Prevention of Cruelty to Animals. It handled the matter promptly. In only two days a truck came and got the team. As we clapped their thin necks in a sad goodbye, I asked the driver what would probably happen to them. He shrugged his shoulders. "They're done. They ain't worth no more than twenty-five dollars. They'll probably end up in a soap factory." To this day, I have asked myself if it wouldn't have been better if we had held onto them and not bought a tractor. But their teeth were worn from age, they couldn't chew properly anymore, and therefore this end was probably for the best.

Our second greatest concern was putting the buildings in proper order. The roof of the henhouse was half rotted, and since this was otherwise a solid structure—originally the retirement house for Grandmother Harnagel—it appeared to be distinctly salvageable. The more of the old shingles I ripped off, however, the more it was clear that the whole roof frame had to be renewed. Left over from the burnt-down main barn, which had stood in the middle of the field, there were charred but useable roof beams in large numbers. More difficult was acquiring the necessary roof planking. At that time, there was still no corrugated roof sheeting available for purchase. We therefore had to use roofing felt, which could only be applied over a firm underlay of smooth boards. I began to see at that time what a quantity of boards the roof of even a small house swallowed up. I helped myself to the slash boards that lay waiting to rot in great piles at the sawmill. If you nailed them on with the bark-side down, the roof was moderately flat, but this required careful but tedious sorting through boards of dissimilar shape and thickness. By hacking away endlessly, I had the roof far enough along in fourteen days that Rita was able to whitewash the interior and build the nests for the chickens. As the first small flock we bought two dozen Rhode Island Reds at one of the weekly auctions in Perryville. These big copper-colored animals, or the dark gray, black-banded Plymouth Rocks, are far more suitable for rough areas than the better laying leghorns.[2] The first reason for this is that they can protect themselves better against the egg-stealing green chicken snakes, the second is that they are not so easily spotted by chicken hawks, and the third is that they forage in the woods better and give more meat. In the beginning, we feared that our young hens wouldn't get started laying at all. Then, when the unmistakable sound of their having fulfilled their responsibility made it clear that they were laying,[3] we gradually became aware that the eggs were being stolen by rats. By creeping up carefully, we were able to catch the rats red-pawed, so to speak. With astounding skill, they were able to sneak the eggs out from under the setting hens and spin away with them to their holes. This meant it was necessary to set out poison, and further, that the dogs had to be protected from eating poisoned rats. On average, every four weeks, Rita and I fabricated "arsenic butter-bread" until the rat population on the farm had been lowered to tolerable, if inescapable, levels. Each time, we had to get through several days of the anguished yowling of the tied-up dogs, but it was nice when the rooster crowed in the morning and when, with favorable winds, answers came from other farms across the river. It was evidence that little by little our farm was becoming like those others, that it was filling itself with life and was slowly healing.

The next job to tackle was the pump house. All that remained here were the board walls. But since we had determined that the pump house was to be the place for the laundry as well, it had to have a roof; in fact, the little house really had to be completely renewed. Some of the old railroad ties out of our supply were perfect for the sills, and

2. *Leghorn* is the anglicized name for Livorno, Italy. Hauser in fact called these chickens *Italiener*, "Italians."
3. That is, the hens cackled.

since they had a length of almost ten feet, I used them for the upright posts as well. For the roof, I tore down the summer kitchen. It was a half-fallen shed that until now had stood, unsightly and shabby, close to the kitchen wing of the house. Almost all American farms in the south have such airy summer kitchens, so that the housewife suffers less in the heat of summer. Earlier, these summer kitchens also served as smokehouses in the winter. In our case, however, our own house was unusually cool in summer, and the smoking of meats has been replaced by modern, if less commendable, methods of preservation, which I, in my inexperience, didn't want to try. Without a twinge of conscience I therefore ripped the summer kitchen down and got for it instead a quite nice and solid pump house.

Little by little it had become so cold that we had to think about stoves. New stoves in the proper sizes, according to the catalogs, cost up to ninety dollars. For us that was unthinkable. But where were the auctions now? More and more we had become accustomed to reading the auction announcements in the weekly *Perryville Gazette*. Also, in the mail and tacked up in Müller's Store in Wittenberg, there were almost always flyers announcing coming auctions. During and immediately after the war, that was by far the best way to cover farm requirements. For a total of twenty-four dollars, we bought two admirable specimens, one of them a Franklin stove that would take logs up to three feet long. This was one of the many inventions of the boundless genius of Benjamin Franklin. There was still a surplus of scrap wood at the sawmill, but there were none of the hardwood logs needed to heat a house overnight. It just so happened, favorably for us, that isolated stacks of oak logs stood around in our woods. These were wedge-shaped pieces, about five or six feet long, that had been left over from the making of barrel staves, split from oak trunks. Toward the end of the war, however, the orders for these staves had been canceled. Since it didn't pay for the Moss Company to haul them away, the agent had left them in my possession. From the nearest pile I hefted these hundred-pound logs to a place that Perfidio could reach. Or rather, I toppled them over the steep slope to where they could tumble close to the road. Unfortunately, they were too long for the Franklin stove; nevertheless, it took no more than ten minutes of labor per day to keep our rooms warm.

Christmas came as a wonderful time, because we were able to fill our first packages for Europe, but a very melancholy time as well, because so many of the people we had thought about could not be reached or were simply no longer to be found among the living. Also, our means were woefully inadequate in comparison with the extent of the need, and it was clear that our shipments wouldn't reach their recipients in time for Christmas, because the postal connections had only just been opened. Since we couldn't slaughter our own swine yet, Rita bought lard from the butcher in Altenburg. Before the war, lard had cost six cents a pound; the postwar price was six times that. These packages therefore ran into money. For the packaging, Rita had found a brilliant method: She poured the melted fat into empty coffee cans and sealed the cans airtight with melted lead.[4] Coffee was the second focus of our packages; then came sugar, which at the time was still rationed in the United States and therefore very hard to get. In the Fischer's General Store in Altenburg, Rita declared plainly to the owner that she wasn't used to buying sugar a pound at a time. She had to have a hundred pounds for her packages to Europe, and she gave him her whole ration card as her bond.[5] She got the sugar. Altenburgers had a sensitivity for the European situation. What was accomplished by way of assistance in that first postwar winter, considering the poverty of the small farmers here, was extraordinary. In Wittenberg, with its 89 residents, $278 was collected; Altenburg, with 278 residents, brought in $1,400.

When the packages were filled with a little chocolate and Rita's home-canned marmalade, I carried them the five miles to the post office in a shoulder sack, because our road was now almost impassable.[6] I had mastered the technique of balancing on the rails now,

4. It is hard to imagine how she may have done this, and even if the can's contents weren't burnt by the molten metal, one must speculate on how very dangerous these lead-sealed cans would be to the health of the consumers of their contents.

5. Rationing and ration stamps were observed, on the whole. Often, however, a rationed commodity that was scarce in one place was plentiful in others. Country stores and the farm people they catered to apparently had sugar in abundance for the purposes of canning fruits.

6. The distances Hauser gives from the farm to Wittenberg are incorrect. He said here that he had to carry this sack eight kilometers—just short of

so that I didn't have to stumble the long way over the ties. Even so, since there was shopping for the farm as well, the load of thirty to forty pounds each way didn't fall easily to me. I often envied the track workers, who whooshed by me on their railcars. It was strictly forbidden, though, for them to take anyone with them. Of all of our Christmas packages, barely half arrived. An enormous number must have been stolen in Europe at that time, and unfortunately postage was sinfully high, so that a ten-pound package to Germany cost three dollars. Only after a year did the American Post Office reduce its rates for European help, and thanks to the slowly developing help organizations like CARE, deliveries became more reliable.

On Christmas Eve, we lit our little tree, its candles stuck onto clothespins, another Rita-Invention. Then we walked hand in hand over the hard-frozen, crunching grass to the Mississippi, where ice floes wandered by unendingly. There we sat on a stone on the bank under the cold-starred sky, listening to the muffled bumping of the ice floes and thinking about the people in Europe, until the cold gripped our hearts: "Christmas over there must be more like this icy stream than like warm candlelight," Rita said, finally, and suddenly we knew what had driven us out of our warm room and toward the river: the wish to be spiritually near the people over there.

On our way back, to our great surprise, we saw the blinking of a car's headlights in front of the house. It was a wonderful new car, which, thanks to tire chains and hard-frozen ground, had worked its way through to us. Its only occupant now stood on our veranda, and he disclosed himself to be an agent of the FBI, the American federal police. In the course of the war we had experienced the visits of such gentlemen many times, mostly young men of unusually good upbringing and high intelligence. It only astounded us that this one had come on Christmas Eve. But in America the holiday begins on Christmas Day, so that Christmas Eve is considered a workday, and for the FBI it appeared to be quite a long one. We invited the gentleman into the house, and with a cup of coffee he came out with

---

five miles. The actual distance from the house to the post office via the road is 3.4 miles. In this case, Hauser also said he went by way of the Frisco tracks, which lessens the distance to less than three miles.

his concern: A fisherman from Grand Tower had reported that we were possible German spies.

That was new to me, and I asked what the report was based upon. It seemed the fisherman had observed that I had built an observation stand in a great tall tree on the bank, and that all summer I had sat up there and made notes, probably of the number and cargo of the barge tows passing by, and the corvettes that they were building in Cincinnati . . . [7]

Now I understood. The fisherman in question, an old man, a peddler in his main occupation, had been jealous of Rita's fishing success throughout the early summer while we still lived in the doctor's little house. Repeatedly, before day and dew, we had caught him stealing from Rita's fishing creel. Finally, I found the matter just too dumb, and I had roundly declared to him that I would shoot a hole in his boat if I caught him stealing again.[8] In addition, I was a writer, and he could see what crimes I had committed in that tree anytime he liked, and to my knowledge, the war in Europe had already ended in May.

The young official laughed heartily. He had already thought something like that would be the case. With a gesture toward the Christmas tree, he noted that it didn't exactly look like a nest of spies here. He had also asked the postmaster and others in the village about us, and the information had been favorable.

So we sat for another quarter of an hour and philosophized on the stupidity of people who make life so mutually and unnecessarily

7. Whether or not corvettes (armed escort ships) were built in Cincinnati, Hauser could not have spied on them, as the mouth of the Ohio is well below the farm. It is a fact that eighteen U.S. Navy submarines were built at Manitowoc, Wisconsin. These boats cruised to Chicago under their own power and then were placed on barges, where they were taken through the Chicago River and the ship canal to the Illinois River, thence down the Mississippi to New Orleans for final fitting-out. Huc Hauser believed that the informant, if he was concerned about anything real, worried about Hauser seeing and describing these submarines.

8. Henry Scholl keenly remembers this incident, or at least the local gossip about it. He also remembers the name of the complainant, a man named Hoffmann, from Grand Tower. "He was stealing fish from the Hausers' lines," said Mr. Scholl, "and Hauser ran him off. He must have had a grudge."

difficult for themselves. We took our departure as friends, and so occurred this last visit from the FBI that we received in the United States—very different from a corresponding visit from the Gestapo.

At New Year's, the day of reckoning, we counted up our cash and discovered that our bank account was gravely close to exhaustion. That was highly disagreeable, since if the farm was to become something, by spring I would need at minimum a tractor, a plow, hoes, a cultivator, farm wagons, a grain seeder, seeds, and every other sort of thing. On a sharp frosty day, when the ground was once again frozen hard enough, I drove with a heavy heart to the Bank of Altenburg to talk about getting credit. In rural America, even this occurs very differently than in a German bank. The portal through which one enters into the all-holy leads without ado past a bellowing iron stove to the seat of the *Herr Bankdirektor.* I had never seen him before, but he seemed to know me perfectly well. He shoved me a chair and asked, "Well, Henry, you have an awfully long face for such a fine winter's day. What kind of cares do you have in your heart?"[9]

After such an opening, it wasn't hard to march right in with my concern. "Hm, hm," rumbled the old man. "Let me just see Henry's account summary," he called over to a cashier. After a short examination, he clapped the folder shut. "Yes, so it's the farm. You paid for it in cash, yes? Nice, nice. The rest of your money seems to have gone for purchases. Household goods, tools, yes, you used it well, I can understand that. You haven't wasted much, as far as I can see. That's not bad. And now you want a tractor and the things you need for farming?"

Yes, that's what I wanted.

"How much do you need?"

"Twelve hundred dollars. I reckon that would be the minimum."

"Hm, hm, too bad it's more than a thousand dollars. Up to a thousand, we could have given it to you with no other requirements. For this, though, we need a board meeting. Look in on us later in the week. I think you will have your money then. By the way, there's something wrong with the coalerator on my car, and the garageman

---

9. Hauser apparently had seen him before. He was Eugene Boehme, the loan officer mentioned as "Oyshayn" early in the narrative.

The Bank of Altenburg, 1998, as Hauser knew it. It has since been replaced by a modern building. *Photo by Curt Poulton.*

can't fix it. The fellow's forgotten all his German. He says he can't understand me. I told him everything correctly, didn't I?"

As quick as a flash it went through my head that *carburetor*, the English word for *Vergaser*, would translate, word for word in German, as "ver-coal-er." I cleared up the misunderstanding, and the old man and all three of the bank personnel shook with laughter.[10]

"There you see it again! A hundred years ago, when we emigrated, there were no cars, and now we're missing all the new German words for the technical nomenclature. Henry, you should give lessons in the Altenburg Garage, on what auto parts are really called in German."

That I did, and when the new big county road grader came in for repair, though I really didn't know what a road grader was called in

10. It seems obvious from this story that the entire exchange between Boehme and Hauser was conducted in German. As Hauser predicted, today only the older people in the region speak even isolated phrases in German; the lingua franca in Altenburg is solely English.

German, I immediately thought of a carpenter's plane. I rapped on the machine with my knuckles and said:

"This is a *Straßenhobel* (this is a street-plane)."

All the farmers and mechanics roared with laughter, and "this is a street-plane" became a household phrase in the area.

The main thing was that by the weekend I got my twelve-hundred-dollar credit without condition. And what was best about it was that I got it "on character." That means, simply, "on good repute," so that I didn't have to put up the farm as security, as I had feared. Now I could look to spring with a bit more confidence.

Through our contact with the bank we came into somewhat closer contact with the community, even though we didn't go to church in Altenburg. The distance was simply too great, and gasoline was still rationed. Altenburg, Wittenberg, and the other Lutheran congregations in the area lived in a happy past that had almost completely died out in the rest of America. Their pastors had unquestioned leadership, and in secular matters, too, so that the old and established people turned all their disputes over to their "shepherds of the soul" for decisions.[11] Among them it was the same as a sin to go to a lawyer. In this time of deprivation in Europe, it often happened that conniving Germans, most of them not the neediest, searched maps of America for German place-names, and then sent letters to them begging for aid. In the German towns on the Mississippi, such letters were given to the local pastors to deal with, and to the best of my knowledge no supplicant was ever ignored. Also, the synod in St. Louis was the first to dispatch clergy to Germany, whose letters brought about a growing stream of relief.[12]

The most astounding of all these matters of charity was that the village of Altenburg, on the event of its centennial celebration, freely and openly confessed the sins of its first pastorate. Its people emigrated in 1836 under the leadership of this clergyman and naturally

11. Hauser used the common vernacular German term for "clergy," *See-lenhirten*, "shepherds of the soul," here.

12. This refers to the Missouri Synod of the Lutheran Church, which established Concordia Seminary. We may probably infer that Hauser and his family were Lutherans. In and around Altenburg and Wittenberg there would have been many.

turned over to him control of the community funds.[13] When their land on the Mississippi had been purchased, however, the pastor absconded with the rest of the money. With twenty thousand dollars in gold in a pushcart, the disloyal fellow, who was also suspected of being less than chaste in dealing with some of his female parishioners, fled to the Wittenberg ferry. The indignant congregation caught him, however, took back the gold, and cursed and hounded him in disgrace to the Illinois side. He was presumed to have later fallen in the Civil War.

In February, Kitty and Holla dragged a whole nest of young rabbits home from one of their hunting expeditions. Theirs was the purest kind of curiosity and playfulness; the dogs didn't want to do anything at all harmful to the little animals, but they took them into their muzzles, tossed them up in the air, and caught them again. Two or three times we had to grab up the poor creatures, unharmed, and put them in a box under the stove to get warm and recover. Rita's hands flew, warming milk and feeding it to them with an eyedropper in a difficult attempt at nourishment. Two of them died, but the third recovered quickly and, after a few days, played the little man, with his tiny forepaws flailing in the air in impatience when Rita came with the pipette. We called him "PeeWee," because he was so little. He lived in constant danger, because he ran along behind Rita's every step.

In a short time, nothing was sacred to him, neither table nor bed. At lunch he would inform us of his presence by leaving little balls, which we tactfully called "caviar," on our plates. In the evenings, the dogs were brought inside so they could learn how to handle small, helpless bunnies with tenderness and care. Kitty and Holla understood without further instruction that PeeWee's person was sacrosanct. Quietly wagging their tails the clever animals lay down to give PeeWee the opportunity to get to trust them. It wasn't long until something happened that showed us overwhelmingly what life in Paradise would be like. Half-creeping and curious, with his little nose snuffling, the dwarf rabbit crept up to Holla, who lay near him,

13. Actually, the Saxon Lutherans arrived in St. Louis in 1838 and in Perry County in 1839.

and tried to suck at her nipple. Deeply apprehensive, we watched the dog raise her forepaw, which was bigger than PeeWee, but she laid it so softly over the body of her adoptive child that PeeWee kept on suckling. Jealous Kitty wouldn't rest until she had been given the same favor, and from then on our evening reading hour was over, because the drama of these three intimately friendly creatures was much too fascinating. PeeWee played hide-and-seek and let himself be dragged out of dark corners in their mouths. PeeWee raced in great leaps onto the bed and sprang from there onto the neck of one of the slumbering dogs. PeeWee nibbled at his milk toast, and when the dogs tried to get some of it, he hit them on their big noses with his tiny front paws so hard that they ran away with their tails between their legs.

Once something fearful happened. PeeWee was long since weaned, but his favorite food was still milk out of the eyedropper. Meanwhile, he had grown quite powerful teeth, and so it happened that the glass pipette broke, and he smoothly swallowed a couple of slivers from it. That night we didn't sleep. We reckoned that PeeWee would surely die. We fluttered around his tiny body for three days, while he remained fresh and happy. Then we had no more doubt: PeeWee, the miraculously rescued bunny rabbit, was in reality a charmed prince.

At the end of February, one of the grandiose Mississippi storms came in from the south. The whole sky was in flames; our roof groaned under the drumming rage of the cloudburst. Around midnight the storm subsided, and after a minute of meaningful silence, the bullfrogs awoke from their winter's sleep and burst into their concert, roaring with the resonance of elk in the rutting season. When dawn came and we stepped out onto the veranda, all around us the woods, until now gray, shimmered in the brightest of young green. Overnight it had come—the Mississippi spring with its sudden primitive power.

The Tractor

In this second year on our farm we knew a bit more about the nature all around us. In Wittenberg there was an old plant collector who came by us from time to time, and from whom I learned a great deal. We learned the most, though, from the old negro professor, George Washington Carver, of the Tuskeegee Institute, with whom I had corresponded from my first farm in New York State and had traded dried mushrooms. This wonderful old man had been born a slave. Through his scientific work he had revolutionized the agriculture of the entire south, and had done more for his race than any other. Just before his death he had sent us a little book about American wild vegetables and wild fruit. It was only then that I first discovered word for word just how well one could live "off the fat of the land." Rita adopted his methods for preserving berries without canning apparatus or glass jars. One simply mashed the fruit to a flat paste, and let it dry in the sun. I found whole fields of "lamb's lettuce," and for a spinachlike spring vegetable there was "poke," a weed sometimes as tall as a man, from which only the youngest growth is flavorful and wholesome. Carver's book told me that there were no fewer than twelve different sorts of walnut trees on our farm. In the middle of the woods grew peach and pomegranate trees which I freed from choking vines with my axe so that we could look forward to a rich harvest. The land was far more a paradise than I had originally known.[1]

At the time our valley came into full bloom, when nut trees and vines especially perfumed it with indescribable delicacy, our fisherman friend Leo Harris visited us with his little daughter, Naomi. He had split up with his companion Bill, "because he drank himself dumb and stupid," and now Naomi steered his boat. Leo was a

---

1. The second year on the farm was 1946. In this paragraph, the wild leafy plant Hauser refers to is probably lamb's-quarter, a common weed that can be used as a pot herb as well as a salad green. Another wild edible, pokeweed, is also known in the south as poke salat. Its leaves must be cooked thoroughly to be edible. True pomegranates probably did not grow on his farm, however.

widower. One of his sons was a crewman for the Federal Barge Line, another son was a soldier in the Pacific, and Gale, the youngest, at sixteen, was already a highway captain, piloting his truck over the eighteen-hundred-mile stretch between St. Louis and Los Angeles. Thirteen-year-old Naomi was therefore Leo's steersman and housemother at once and truly didn't have an easy lot. The girl was very nearsighted, but her father had no money to buy her glasses. The school doctor had a pair of glasses made for her, but her father ordered her to give them back: "For my part there shouldn't be a government, and no government support." Leo had a good shot of Indian blood, and perhaps because of it, a wild pride, a fanaticism about independence. All his experience as a fisherman came to nothing, but that was not his fault. Leo Harris had only half a lung and simply didn't have the strength to haul in the heavy nets. Naomi went around literally in rags. She had apparently never been acquainted with a bathtub, and since she didn't even have a comb, her hair was matted like felt. While Leo and I sat on the veranda and attended to men's business, Rita did a little something for Naomi. Washed, combed, and done up in one of Rita's quickly altered summer dresses, the little girl appeared before us beaming. Her heartrending gratefulness repaid Rita in full.[2]

Encouraged by all this, Leo slowly began to come out with his concern. During the first years of the war, he had operated an illegal still on our farm. It had gone very well, and there had been no danger, because he had burned only smokeless charcoal. Unfortunately, last year there had been a mudslide in the gulch where he had it hidden, and the copper kettle, with fifty gallons of whiskey still in it, lay buried under fifteen feet of earth. He didn't have the strength to dig the thing out on his own. What he offered me was no more and no less than a partnership. In his plan, we would secretly set the apparatus up again. There was a lot of profit to be made in it.

Private distillation of spirits is a crime in America. Time in the penitentiary is the price for it. Also, I had never thought to use my land for any sort of hideout that gathers all sorts of trash around it. So, as carefully and tactfully as possible, I turned the offer down

2. Hauser's portrait of Naomi Harris is less sanguine than that of Huc Hauser. Huc remembers the girl as being "very pretty!"

smartly. The manner in which Leo took this spoke for his character. "I basically didn't expect anything else from you," he said. "It doesn't matter. I'll build a new setup over on Pig Island, in the middle of the river. Not even dead animals show up there. So, we'll stay good friends, and if I can help you in any way, I'm ready day or night."[3]

Leo was an excellent mechanic, so by way of answer I told him about my most difficult commitment for this spring: the purchase of a tractor. At that time, tractors were extraordinarily scarce. Tractor production had been halted for almost the whole of the war, and the majority of the postwar production went to Europe. Clever speculators turned this scarcity to their own advantage by buying up every make and model in fall and winter, and in the spring, when farmers most needed them, these sly fellows held auctions in the towns of the farming districts. Their prices were of course sharply raised. With Leo I drove from auction to auction, always horrified by the prices, which would have swallowed up just about all of my credit. And when we found an inexpensive hayrake or an old but useful grain seeder and towed it home behind Perfidio, the most important item was always missing—the power source.

One day, Leo came walking up to the house from the river, as usual, with Naomi in his wake, and in a very mysterious but broad manner he confided:

"There is a colored hauling contractor in Grand Tower who is about to go bankrupt. He has a nice little John Deere tractor, a two-cylinder, and he needs cash for it, urgently, today yet. Would you like to go over there with me now?"

And how I would! I was so sure that something was going to come of this, and so concerned that I would buy foolishly out of pure eagerness, that I asked Rita to come with us. It was our first visit to the negro quarter of Grand Tower, where Leo lived, incidentally, and I have to say that I had never seen such filth and poverty. This area had not yet been cleaned up after the great flood of last spring. The tumbledown wooden huts still had mud crusts halfway up to their roofs. People, chickens, and pigs lived together in the same room. We were immediately almost ready to go back, but Leo stopped me

3. What Leo calls "Pig Island," Hauser later calls "Treasure Island." U.S.G.S. topographical maps do not name it at all.

and wisely prompted: "Be careful, don't say anything about why you're here. Tell them you want to buy a farm wagon. The man doesn't know what a peach of a tractor he has in his shed. Naomi, you take Rita to our house. This is men's business, it has nothing to do with you."

That women were not to be involved in important things was an Americanism of the old style, and the half-bankrupt hauling company we nonchalantly strolled through was of the old style, too. The nearly collapsed shed was so stuffed with dust-covered wagons and ancient, rusty vehicles that it looked impossible to penetrate its depths, and the barefoot, gray-haired old negro who stood by the blacksmith's forge seemed more to belong in the time of slavery than in our century.

Wordlessly, Leo took the hammer and tongs out of his hands and in a few minutes finished welding the wheel rim that the old man had doctored around with up till then. "You sure have a good hand, Leo," the man said, admiringly. And then we came slowly to the point. "Farm wagons?" he said, "No." The kind I wanted, with steel wheels and roller bearings, he didn't have. But he had plenty of old trucks, out of which he suggested I could make a first-class wagon.

Did he have anything good to sell? Leo pointed out, "The gent here bought a farm over in Missouri. He has cash in his pocket. Are you going to send him back home empty-handed?"

The old man scratched himself sadly, first his behind, then his head. "Well, I have the tractor, but I don't want to sell it. I want to use it to build up that ditch by the river."

"Phooey! That gully will get flooded again this year just like last year and every year. You'll never build it up. You're a little bit too old for that, as well. Let him see the tractor. I doubt the thing will still run, anyway."

"Run? Man, with that tractor I can pull your whole house from under you! It's got power! Power, I tell you!"

Through spiderwebs and over trucks and graders we made our way to the back, and there, small, compact, and frog-green, "He" stood on proud, fat rubber tires. I fell in love with him at first sight.

"Man, oh, man, what have you let this poor thing come to!" shouted Leo. "It's sitting right under that big hole in the roof. I'll bet there's water in the gas tank and all of the pedals are rusted tight."

By way of an answer, the old man pushed in the throttle and swung the flywheel a couple of times; a little white cloud puffed out of the chimneylike exhaust pipe, and in the next moment the shed was filled with good, hard engine sounds. "Is this a tractor, or isn't it? Isn't it worth at least a thousand dollars?"

Leo spewed tobacco juice contemptuously. "It cost seven-fifty when it was new—ten years ago. It's not worth four hundred today. Think about everything on it that has to be fixed."

The next half hour passed with shrieks and shouts between Leo and the old man that even the droning engine couldn't drown out. Then came the solemn moment, when the old man and I squeezed each other's hands with all our might, and Leo smacked our fists and said, "Done!" For six hundred dollars, the John Deere was mine. "A gift," murmured Leo in my ear, "he gave it to you!"

After a couple of boards were broken out of the shed wall, we only had to push lightly and the wall fell flat. Then we just drove the tractor out. Leo triumphantly drove it to the garage first, where we filled it with fuel, and then went to his little house, where Naomi and Rita came running up to us.

"Leo," I said, "you did that business really well, and it's right that such matters have to be celebrated. But just tell me, now, how are we going to get the thing across the river? The Wittenberg Ferry doesn't exist any longer, and the bridge at the Cape is at least thirty miles away. It would take days to get it to my place."

"Days?" Out of his inexhaustible supply, Leo laid another chewing tobacco oyster on the ground. "In a half hour we'll have the thing over there. Want to bet?"

"Five against one," I said quickly, because when Americans bet, the matter is serious, "But how are you going to do it?"

"Very simple. We'll roll it onto my boat."

"You're crazy, Leo. That thing weighs over a ton!"

"So? I've carried over two tons in my boat."

"Impossible! Think about the balance. As sure as anything the boat will tip over!"

Compassionately, Leo laid his hand on my shoulder. "You'll see. It will go perfectly. We'll take Black Bill and his boat with us. Even if he is a drunk, he's the best sailor on the Mississippi, and he's as strong as an ox, too."

With a heavy heart I pressed the money for the party drinks into Rita's hand and followed the tracks of my tractor down to the boat landing. "It's a shame," I thought. "Such a pretty little tractor, and now it will most certainly sink, and probably us along with it."

At the landing, with the involvement of the whole community, both boats were lashed together. Children dragged up heavy planks. In a few minutes, a halfway solid platform had been built on the two boats. My tractor, though, shivered on its rubber tires on the shore. It seemed just as indisposed about this voyage as I was. Carefully, Leo let the machine slide down the muddy slope. The boats bobbed under the load. For seconds it looked as if they would break apart and the tractor would fall between them, but again it went well. A bit pale, we took our places between the wheels. Rita held her hand overboard and showed me secretly how we had barely three fingers freeboard. The matter wasn't to be stopped, though. Prosper or perish, we were bound to our tractor.

Never had I so carefully watched for floating tree trunks, and never had Mississippi eddies looked so ominous as this time. The worst moment came when we neared our shore and the boats had to be turned upstream. Leo's boat took on substantial water. The river was at half high water, and even though I prayed to the heavens for Leo to land at the doctor's little house on the peninsula, where it was most placid, no, he insisted on landing right by the farm. With unbelievable skill, he and Black Bill steered their boats up the creek and under the railroad bridge. They couldn't possibly have expected that this would work, but the tractor glided under the bridge with a finger's breadth of room and now, even though it was also flooded, we were at our own land. Carefully squeezed between the trees, the boats drove up to the spot where our farm road crossed the creekbed. There, we let them simply drift up, and while Bill and Leo held the rudders I drove the tractor over the creaking boards onto my own dry land. I can't describe the feelings of thankfulness and luck I felt.[4]

As if Heaven had not done enough to bless this day, ahead of my tractor wheels I suddenly saw two miniature tanks strolling over the

4. It would be easy to assume that the ferrying of the tractor across the Mississippi River on two small boats was fantasy, but Henry Scholl reported that this account is absolutely true.

field—turtles of a sort I had never seen before. At my yell, Bill and
Leo came running. "Snapping turtles!" they shouted excitedly. "And
whoppers! The big one there must weigh thirty pounds." They both
pulled out their pocket knives and the little round whetstones that
every fisherman carried in his pocket, spit on them, and began to
whet the blades vigorously.

"Can you eat them?" I asked innocently.

"And how! They are even better than the diamondback turtles
down in Louisiana. Run, Bill, and bring a couple of sticks."

The process of slaughtering them was not exactly pretty, but it
was altogether interesting. Bill and Leo held the sticks out in front
of the hissing animals. Without a moment's hesitation, the beasts
snapped onto them with the power of a pair of tongs in the hands of
a strong man.

"If those were our fingers," Leo said, "they would have bitten
right through."

Now Leo bent the stick backwards over the shell of the animal,
and since the turtle wouldn't think of letting go, the long, white
underside of its neck turned upward. When Bill completed the same
maneuver nearby, each throat was cut through lightning fast.

"Ugh! That's awful," said Rita, and she flew with long strides
toward the farmhouse. The view was quite bloody, and even though
their heads hung on only by a flap of skin, the turtles' elephantine
legs still rowed wildly and with such force that one of them sliced
through my boot with a claw. The wriggling bodies were carried
by their tails to the house, where Leo asked for a washtub and set
about disassembling them. First, the upper and lower shells were
separated with an axe. Then, a knife was slid along the inside of
the shell to loosen the tough skin and sinews from the armor plates.
The opened bodies showed the strangest construction I had ever
seen in any animal. The musculature was limited almost entirely
to the extremities. There were neither back nor stomach muscles,
and all of the inner organs swam in pools of a crystalline fluid.
To pull the unbelievably tough skin off of the legs we used pli-
ers, and it took a good half hour to do it. In the end, the curiously shaped pieces of soft coral-red flesh half filled the washtub.
The two animals may have given up well over twenty-five pounds
of meat.

Leo supervised the cooking himself, and there was a banquet so delicious that I doubt any Parisian restaurant could have surpassed it. The delicacy surpassed any fowl, and the only disadvantage to the meat is that it is so rich that the diner bites off more than he can chew. Bill told us that it was forbidden on the old slave farms to give the negroes turtle meat more than twice a week, because they would get too cocky. I didn't doubt it.

Then we sat on the veranda, passing the bottle and scrutinizing the tractor profoundly, until the darkness finally swallowed the contours of the machine. Bill and Leo spoke wisely of all the tricks of cultivation and plowing and argued about the thickness of the roots that this tractor could cut through, and which fields would be best for planting buckwheat and corn. Leo had been a tractor driver on the farm of a rich banker from St. Louis for many years. He had had it good there, but then one day he just said, "The devil with this job."

"But why, Leo?"

"Oh, you know, these rich people can kill your spirit. Every week the guy drove across the field in his Cadillac and told me long stories about what awful business problems he had. It finally got too much for me. I don't think rich people have the right to sermonize to the poor about how their lives are so plagued."

The later the evening moved on, and the more the golden surface of the second whiskey bottle sank, the stronger the conflict within Black Bill grew. He confessed quite frankly that he hated the white race from the ground up, present company excepted, of course. Negroes had no chance in the United States. It would be best if the whole black population went back to Africa, an exodus like that of the Children of Israel. He had a good mind to become a seaman just so he could see Africa and if possible to live there. But the old slave blood awoke along with the hate as well. While he had addressed us by our names before, now he mumbled "boss," and "boss lady," and shriveled more and more into himself until he finally threw his woolly gray head onto the table and howled in misery.[5] Leo gathered

5. Henry Scholl stated that he thought the character of Black Bill may have been based on a man from Grand Tower named Primer Adams. After reading this description of the man's anger at white people, he said, "Then I am sure it was him."

him up, and even though I offered beds for the night, concerned about how the two of them would find their way across the river and home, they stumbled into the darkness.

I needn't have worried myself so, because the next morning I found the two of them peacefully sleeping in front of Perfidio. After Black Bill had slipped and fallen in the mud a few times, they had given up the race. After breakfast, Leo took my tools and pulled the cylinder head off the tractor. "Like new from the factory," he said admiringly. "Doesn't even need new piston rings." Then, reflectively, he scraped away the carbon with his pocketknife, cleaned the spark plug, screwed the whole thing together again, clapped the machine on its green back, and said, thereupon, "Good for two years."

And so ended the twenty-four hours of the second greatest day on the farm.

Seed and Forest Fire

I built a skid out of an old sleigh, threw the plow into it, the only one I had, chained it to the tractor, and drove proudly to my first field. The six-acre patch was west of the house. In the meantime, Rita had become an outstanding overland driver. She had a couple of test drives on the tractor behind her as well. So, Rita would sit on the tractor and I would guide the plow. Or at least so we had planned . . .

The reality was something else. At first, the tongue was installed backward, and, as I observed the first furrow, it was not only crooked, but the ground had barely been peeled back. Even worse, the tough layer of weeds that had grown up on the fallow land caused the clods to rush back into their old position as if pulled by rubber bands. It was like witchcraft, as if no plow had come that way at all.

Nearing the other end, I lengthened the chain that fastened the tractor to the plow, so that the plowshare went deeper. For twenty paces everything went well, but then the blade dug in deeper and deeper and I heard the motor working harder and harder. Since I had thrown myself over the plow handles with all my weight in order to keep the plow up in front, I wasn't in a position to see what was ahead of it. Suddenly I heard a fearful cry and saw the tractor rear up, so that Rita came within a hair of falling out of the seat. The radiator stood almost vertical against the sky, the front wheels spun in empty air, and since Rita had forgotten to shut off the motor in her initial shock, the machine stood still but its rear wheels milled in deeper and deeper. The plow had gotten stuck and had anchored the tractor, so to speak, and there was nothing left for the machine to do but to rear itself up this way.

After I had very carefully freed the plow, the same drama repeated itself in the next hundred yards at least five times. Big stones lay hidden in the ground, and every time the plow hit one, I felt a sinew-ripping tug on my arms and the tractor reared up like a frightened horse. This had become dangerous; we weren't moving, and it had really come down to my wife's life, because sometimes the machine didn't quite stop, and a backwards tug by the plow could have ended very badly. When we observed the work around noon, fully exhausted and sweaty, we saw that we had drawn only two furrows, and they looked as if they had been rooted up by a herd of wild pigs.

"It can't go on this way anymore," said Rita. "This afternoon, you drive the tractor and I'll plow."

This did not suit me at all, but what could I do? Behind the plow, my wife was at least not in danger for her life. I also imagined that I could control the rearing up of the tractor better than she.

The afternoon went just as agonizingly as the morning. After many tries, the plow was finally adjusted so that it held its depth moderately well, but again and again it drove us off to the edge of the field, either because of the massive roots or the accursed stones. I couldn't control the tractor's rearing any better than Rita. Every time it happened, I felt in my heart the tearing yank that had gone through her arms, and when I could look back, it was a torment to see her tender little form behind the plow, her dust-covered face cramped with tension, the tendons in her arms distended from the effort. If any proper American farmer had seen us so, he would have considered me a completely heartless and horrible person, and to me he would have been absolutely right. When we looked at the day's work that evening, we saw a plowed strip barely fifteen yards wide, and it was easy to see that at this rate we would never have so much as a single green shoot. Naturally, there was a suitable John Deere plow for our tractor, but it cost at least a hundred and fifty dollars, and in the year 1946 it was available only to the oldest and best customers.

To make the story short, it took a full week to plow this six acres. We could almost have done as well with a spade in the same time. Our mood was of the blackest doubt. Even Leo, who came over on the weekend, shook his head. But, as usual, Leo had an idea. There was a farmer over there in Grand Tower who had a "riding plow," a plow with a seat that you drove, that had stood in a field for years. He apparently didn't need it and might well sell it cheaply. Within the same hour I drove over with Leo, bought the plow for twelve dollars, and had the smith weld a new edge on the share and sharpen it.

With this apparatus the matter went noticeably better. Admittedly, I had to climb down from the tractor at the end of each furrow, pull the lever on the plow, heave the share out of the earth, move the tractor into position for the next furrow, and then climb back down to set the plow depth. The jumping on and off became a habit, though, because the plow gathered so many roots and twigs from the weed-grown field that it had to be cleaned every couple of minutes. I did better in the morning, first, because it was moderately cool then, and second, because Rita rode with me on the plow and shoved the masses of weeds off the share with a hoe. The best part of my field was overgrown with four- and five-year-old cottonwoods. As fast-growing as they are, and in this warm climate, most of them were as thick as an arm. Since I was practically plowing in a young forest, this meant that I had to stay on the tractor to see where I was driving, and it meant as well that I had to give it more gas at every tree so that the plow would cut through the roots. Rita sat surrounded by falling young trees, and as well as she could she warded off the lashing branches with both hands. How these little trees were able to bore through the spokes of both tractor wheels is a puzzle to me, but thanks to the pure malice of the things, they managed it often. Every time this happened, the whole apparatus came to a halt and had to be driven backwards with great trouble until the spiteful trunks could be removed. Toward evening, I uncoupled the plow and hooked up the skid to haul away the trees. Masses of undergrowth stood at the edges of the field, but we didn't dare burn it because the weather was dry.

In the third week I discovered a trick that any old farmer could have told me, but there were simply none there to teach me. I started by plowing a strip about fifteen yards wide, and then plowed an oval around the middle. This allowed a continuous effort. It was no longer necessary to pull the plow out at the end of a furrow and reset it, because the furrow was endless. Admittedly, the shape of my field grew more and more oval in form and its edges weren't plowed cleanly. That didn't matter much, because we had more land than we could work, anyway. Never before had I looked forward to the arrival of my son as I did this year. Apart from woodchucks and countless rabbits, I grubbed up dozens of snakes. Most were harmless blacksnakes that one rightly called "the farmer's friend" because they are outstanding mouse hunters. My feelings about the rattlesnakes and copperheads were very different, and I made a habit of carrying the shotgun with me on the tractor. That spring I shot eleven rattlesnakes from the seat of the tractor, and I didn't take the trouble to get down and break the rattles off of their tails.[1] I let them lie there until the plow came back and buried them. In California, rattlesnakes are considered a great delicacy and are served in the best restaurants. We couldn't bring ourselves to try them.

In the course of April, as I had plowed approximately thirty acres, it came time for cultivating, which in comparison was a pure pleasure. Masses of roots had regrown by the time of the second round of hoeing. Nevertheless, it was a joy to feel the summer warmth of the satiny earth now. I began to understand the things that mean happiness to farmers.

Seed that year cost, according to type, ten to eleven dollars per bushel. That equals forty-two to forty-six gold marks per sixty-six pounds. This seed is of course of especially high quality, a hybrid corn that matures in ninety days and is made especially fertile and protected from disease by chemical treatment. Corn is also planted in this rich earth with an economy that one could not imagine in Germany. The rows are planted three feet apart and the grains three feet apart as well.[2] We had no tractor-powered seeder, only an old machine with a shaft for a team of horses that required constant coupling and uncoupling. With a wheel track of twelve feet, it would plant four rows of corn at the same time. The seeds were held in two drums that had a triggering device for control. When Rita heard a simultaneous clicking, everything was fine. If it went silent, we had to stop instantly because it meant the sieve was plugged up. Like old farmers in Europe, we prayed at the end of each day of sowing, asking above all that the flood would pass us over. On our farm, like almost everywhere on the Mississippi, the best fields lay in the alluvium of the river. The hilly land wasn't bad, but it was much less fertile because it didn't receive the fruitful layer of mud that every flood deposited. The flood is therefore simultaneously a blessing and a curse. The Mississippi farmer sows as late as possible, after the spring flood, that is, and it is for this reason that the quick-ripening

1. According to folk legend, rattlesnakes might come alive again if the rattles were not removed.

2. Modern cornfields are planted in the wide rows Hauser mentioned, but individual plants are set much closer together, perhaps a foot apart.

"ninety-day corn" is planted. If the snowmelt in the north is late, and therefore the flood with it, the farmer still has to plant his alluvial fields in the hope that the flood will spare him. Mostly the speculation is thwarted. The first seeds are washed out or the young plants drown. If there is still time, a second planting can be undertaken, but in such cases the farmer is mostly content with the faster ripening buckwheat.

After eight days, when the first sword-shaped leaf broke out of the black earth in the brightest green, we experienced a quite new and a great feeling of happiness. How thankful this earth was; how much more thankful than any other in the world. At sunup and sundown we went to our fields to admire this emerald green in the slanting light, and we could never get our fill of it.

In the beginning of May, two of our sows farrowed, evidently without pain or difficulty. Again, we experienced a miracle; one of our hens, an old one about which I had my doubts, adopted one litter of eleven piglets. The sight was as grotesque as it was moving, when she took the piglets under her wing in the evening and tried to lead them around all day. Now that the corn was up, naturally the swine had to stay in their pen. This meant an increased demand for feed. With more time and more money I could have saved on feed. A mile or so of hog fence would have given the growing swine herd their freedom, but it is always so in agriculture that out of poverty comes need.

Quite different and much better this year was the work in the garden. I built two sufficiently wide gates in the fence so that in one day I had plowed and cultivated the whole garden. We planted seed potatoes in the American style this year, namely, in flat furrows and not covered with earth. Instead, the plants were covered well with "trash," that is, with rotted compost, and only after the seedlings had broken through this layer were they heaped over with earth. In anticipating the harvest, this new method brought easily twice that of the old. The previous fall, I had discovered numerous asparagus plants at the edge of the woods near the house, apparently grown wild. I had dug up a few hundred and stored them in the cellar. So this summer we had our first asparagus patch. We didn't try to cultivate tomato and pepper plants ourselves, because we didn't yet have a hothouse. They came from Sears and Roebuck at only a penny apiece. Here, there was another thing to learn from the Altenburg farmwives: You don't plant them upright, but slantwise, so that the plant's stem almost lies on the ground. This way the young plant is much better protected from the wind, and, above all, it drove down an immeasurable number of additional rootlets. The stem would later grow upright on its own.

At the end of May, our son arrived. I didn't take the car to get him from the station but drove the tractor, instead. For luggage and shopping, I had long since had a big wooden chest hung onto it; now I generally had a 100 percent safety in our transportation, because the machine triumphed over fully nonexistent roads in which any auto would have become stuck fast. In Huc's voluminous luggage, the main item this time was a small gasoline engine, probably from a washing machine. Recovered somewhat from his joy and wonder over the tractor, he bubbled forth how this summer, with the help of this motor, he was going to put our agriculture on an entirely new basis: a) a boat was to be built; b) ditto a motorcycle; and c) the water pump was to be motorized. When I objected that all three projects couldn't be completed with one motor, he countered indignantly, "You just don't understand; that's the best part of my invention: a moveable motor, which can be installed interchangeably in a boat, a car, and so forth, wherever you need it."

At first, however, we had quite different and very difficult concerns. On the afternoon of Huc's first day of vacation, I saw, full of grim premonitions, a black cloud rising over the hills to the south. The boy and the dogs came running quickly out of the woods; "Father, the valley right behind ours is burning. The fire is moving up from the railroad grade. Father, what are we going to do now?"

Wildfires are a yearly occurrence on the Mississippi in the early summer. Occasionally they are started by railroad grade workers, who burn away the bushes on the right-of-way. Often fires are laid by farmers, out of the old belief that burning off the fields makes them more fertile. Most of them are set accidentally, though, by thoughtless hunters, mostly younger people. They want clear paths for shooting, but it is a burden for them to cut through the underbrush. They don't think for a second what sort of mischief their little fires create, often for miles around.

They had to have seen the smoke cloud in Wittenberg, too, because immediately after the boy and I set out to find out the extent of the fire, a couple of old farmers came rattling up in their cars. Just as it is with farmers the whole world over, these men groused profoundly but did nothing. They had plenty of opinions, though: The way the wind stood, the fire would without question move north, climb the escarpment, and probably lay waste to our whole woodland, where there were so many felled tree crowns for nourishment. Only a heavy rain could save us. But the weatherman on the radio had mentioned nothing about rain. The only thing we could do was to plow a strip around the house as quickly as possible. I might be able to save the house that way, but only if the wind didn't freshen, as it probably would.

While they swore that the track gangs were probably at fault, the boy cried out and pointed his hand toward the south; a second smoke column rose threateningly red and black behind the hills.

"Dammit all to hell," roared one of them, "that has to be at old Dick Wilson's! Let's get out of here—otherwise we won't get through and back home!"

In the next few seconds, the motors growled, and the only people we could hope might help us raced up the north hill at full gas. Rita's face was white: "We have to do something. Maybe we can beat out the flames if it's only a brush fire."

Even though I knew how totally hopeless it was, I didn't want to take away her illusion; also, it would be better if she saw it with her own eyes.

"Bring two picks and a shovel," I called to the boy, "and don't forget to fill a couple of canteens—we'll need them."

We followed the railroad grade for a mile and a half, to get around to the back side of the hill where the fire had broken out. Then we could see the valley on the other side of ours. The first couple of hundred yards the woods stood smoking and all the brush was heavily blackened. The fire had clearly moved up from the rail grade. Deeper into the woods, there were small tongues of flame, and only where rotted tree trunks lay curled thick white smoke. "Let's go in and up the north slope," I cried, "so we can get a clearer view!"

Like a condemned man, I still hoped it wasn't so bad; maybe this curse could be lifted from us. Gasping, we climbed three-quarters of a mile up the untouched slope to the highest point, several hundred feet above the river. The lumbermen had cut a road up there along the ridge, and I knew that from that point one could see down into the valley. All around us, nature was agitated: Woodpeckers weren't drumming, but were flying from tree to tree instead; hundreds of hawks screeched; rabbits and squirrels fled, not furtively but with heedless noise, all of them to the north. When we reached the ridge road, all three of us stood as if set in concrete: "My God," whispered Rita. "My God!"

As far as we could see, the fire's front advanced through the valley and up the north slope towards us, a curved battle line a good ten miles in length. The edge of the tongues of flame were still well over five hundred yards from us, but we could already hear the dull roaring. This valley had once been, like ours, fields and meadows. For years, though, the farms there had been abandoned and the young forest had overwhelmed it. Pines exploded into flame with cracks like gunshots. They burned like torches for only a minute, then stood in thousands, their trunks naked and black. A brook, hidden by the undergrowth until now, was visible behind the flame front, its water steaming. I knew it led to the abandoned houses, and by its meanders I could more or less reckon where these houses stood, even though they remained invisible in the smoke. All at once a mighty column of fire shot up over the smoke cloud and we knew that now the houses were burning. In no more than ten minutes the pillar of fire collapsed and left only a column of smoke, twisting and boiling on the spot. Such old dried-out wooden houses burned like matchboxes. It would be just as quick—that we knew—with our own.

We looked at one another and our eyes sank to the laughable tools in our hands; two picks and a shovel against a forest fire of such enormity, dear God! A bitter and ironic smile swept over Rita's mouth. While we had been standing there, the snakes of fire had been creeping through the trees and grass. Many hollow trees in the area began to smoke like chimneys. Inside, they droned dully, as if the trunks had been bass viols, and then quickly the flames blew upward, roaring deeply. The smoke clouds made us cough and the heat dried our skin as we slowly traveled the ridge, following the progress of the fire's offensive. In the luxuriant greening hardwood forest, the fire destroyed only the undergrowth, the leaves from last fall, and all dry wood on the ground. The damage was immeasurable even so, since the trunks were blackened to the height of a man and the bark was glowing. Even if the hardwoods lived through this, the burnt feet of the trees would offer surfaces for insects and rot to attack. This forest would be crippled for years to come. The dance of the flames was eerie in the green darkness of the woods; the white snakes of smoke that wound through the trees were ghostly, and the contrast between the lifeforce of the green dome and the red death hellishly gnawing at it made us shudder.

The tangle of hills in our area is unending. Even though the broader valleys cut down to the Mississippi, they have been cut into laterally by a hundredfold erosion gullies. Nothing is easier than to lose one's way, especially when you can no longer see the sun because of heavy smoke. We lost our direction a couple of times and found ourselves encircled by the ring of fire. To get out faster, we balanced like tightrope walkers on tree trunks that had fallen across gullies, and in general did things that we would never have undertaken in normal human rationality, until we were able to leap over the fire front and away.

We saw our farm finally, like a happy island in the valley floor, and almost cried from joy, from sadness, and from exhaustion. As we drank ourselves full in the pump house we discovered that all three of us looked like chimney sweeps. There was no time to wash, though, and no desire to take time to laugh over our appearances; there were much more important things to do. The fire front was advancing toward us from the south at a speed of about five hundred yards per hour. Since the wind would turn somewhat westerly toward evening, it would also encompass our farm. And when we looked northward, we saw with shock that the fire wall had advanced nearer to us. It is simply not true that forest fires move only with the wind. If the flames find good nourishment on the ground, they will go against the wind.

As fast as we could, we hooked up the wide cultivator to the tractor. With it, we could strip more earth in less time than with the plow. Also, the plow would have gotten stuck too often in the heavily wooded land behind the house. Rita drove the tractor while Huc and I ran ahead with axes to clear a path for the machine. The freshly cut earth would protect the house on the south and west, but the wooded escarpment crowded right up to the yard on the north and east. It climbed to Cemetery Hill and into the old, somewhat dry fruit trees in the garden. Here it was necessary to create an opening before it was too late. In the race with sundown, the bloodiest we ever saw, Huc and I cut down trees, mainly the spruces and firs that were the most prone to fire, while Rita maneuvered back and forth, working the cultivator blades until the thick layer of leaves and needles was covered by a layer of moist earth. Oh! We accomplished so little, so fearfully little, in these last hours of daylight. The glade was barely fifty yards long when it was pitch dark and our poor little tractor had run itself so hard aground into a heavy bit of woods that we had to go and get a winch and a jack.

As we tumbled down the hill, scratched and tattered all over, with pitch-sticky hands, a tall, skinny form appeared like a shadow from the stream. Naturally, it was Leo, the only one who could concern himself about us, because his side of the river wasn't threatened by the fire. We were so exhausted that we let ourselves fall onto the floor of the veranda and for once to smoke a cigarette. The presence of the experienced old man calmed us as usual. After the long silence that was his style, he slowly began to speak. "I could see from down at the river what you were doing up there with the tractor. That won't do anything. You'll have to lay a backfire. But not now; first in the morning. Tonight you're not threatened."

"I'm afraid, Leo," I made this openly known, "I've never ever laid a backfire; I'm afraid we'll lose control of it and our house will burn down."

Leo spit tobacco juice. "You needn't be afraid. There's nothing to it if you understand it. I'll stay with you tonight and as soon as it's light we'll get started."

Once again we climbed up the west slope, with flashlights in hand. Shaking his head, Leo observed the fast-stuck tractor, cranked up the motor, and while we bent the young trees away from the wheels he rocked the machine free with fast shifts from forward to reverse. Nobody else could have done this without a jack and a winch.

Meanwhile, little Naomi had come up from the boat and wordlessly lit a lamp in the kitchen and fired the stove. We were thankful from our hearts, because Rita had given up the last of her strength and lay collapsed over the bed breathing heavily. We will never forget this night. Our farm was surrounded on three sides by a wide arc of fire; slowly and irresistibly the flame front pulled its loop tighter around our valley. Behind the flames was the hillside, covered with the huge torches formed from hollow trees. From time to time we perceived a wild crackling and saw fountains of sparks when a pine grove was attacked. In a dozen places, where in previous years sawmills had left huge piles of sawdust and scrap wood, rose huge columns of glowing heat. From down on the river ghosted the spotlight of a towboat, and from the south sheet lightning flickered spectrally through the smoke wall. Over our pasture, millions of fireflies danced, and through the silence of the night we heard the fire bell ring in faraway Altenburg. Wildness and danger and grandiose beauty were so magically mixed that words cannot describe it.

Often in this night Leo and I climbed with slow strides up the hills—Leo could go no faster because of his missing lung—quick to the north, quick to the south, to check the progress of the fire's assault. Since there wasn't a breath of air, the march of the flames had slowed; perhaps a hundred yards an hour, I reckoned. When we came home I took a look into the bedroom, where Rita and Naomi, both fully dressed, slept the sleep of the exhausted. Our boy lay rolled up between the two dogs. He had naturally wanted to keep watch, but was overwhelmed. He mumbled in his sleep with his legs twitching, and Kitty and Holla also groaned in their dog dreams. Most of the time, though, we sat leaning against the house wall, smoking continuously, in almost complete silence, except that I asked again and again, "Leo, don't you think the sheet lightning from the south is getting closer? Don't you think we're still going to get rain?" When Leo's only answer was to shake his head and spit, I prayed as I had never prayed before, that Heaven would want to give us rain. At last I must have gone to sleep, because I awoke to the smell of the coffee cup that Rita held under my nose.

We started the backfire with the first light of dawn. Fortunately, our five gallon gas can was filled, and we also had five gallons of diesel fuel at hand. Huc chopped up big cedar branches for fire brooms, while Leo and I dragged up dead wood to give the backfire a start. The flames rose with a furious roar. The leaves of the trees on the edge of the fruit grove turned brown, and sparks pattered on the tin roof of our house. We had left Rita and Naomi in the house on purpose, so that if necessary they could save the most valuable of our things. More than once I was to the point, and back again, of giving the order to pack up. Each time, though, Leo managed to bring the fire under control. The whole of the dead, quiet, sun-hot and fire-hot day we kept up this constant effort, so that we didn't concern ourselves with the progress of the forest fire. Then everywhere around the house the tree stumps began to glow with light, and I noticed that it was already night again. We went out with hoes and shovels to cover the glowing heaps of ashes with earth. Our boot soles were smoking, and Huc carried buckets of water in which to cool our feet. Then once again it was pitch dark, and Leo said, "Well, the fire won't get to your house from this side."

And then a miracle happened: Smoke-blackened, red-eyed, and ravenous, we had just sat down at the dinner table when before the kitchen window a snow-white flash of lightning dropped down and the crack of thunder echoed long through the valley. We lowered our knives and forks and sat, wordless and motionless, for fully half a minute.

"That's it," said Leo, finally, "that's what we've waited the whole day for. The rain will come now. God has granted it."

"There it is!" shouted the boy. "Hear it? On the roof!"

Holding our breaths, we followed the fall of the first, heavy drops on the corrugated roof; then there was no more holding back, and we burst outside in joy.

Gigantic lightning and cloudbursts came from the south. With folded hands we stood in the open field, and every time the sheet lightning glowed through the smoke cloud, we shouted with renewed joy. The smoke columns weren't glowing red anymore; they were steaming with white. The ringed tongues of flame around the horizon changed from yellow to red and sank twitching to the ground. The fire front disappeared in a wall of white steam.

Within, we were at first very solemn and still as we thought about what had really happened: From the south, the north, and the west, a forest fire had come within half a mile of our house, everywhere, almost exactly to the boundaries of our farm. With the exception of a few treetops, our land had been spared, while the valleys in the south and west were almost totally ravaged, and in the north the fate of Wittenberg itself hung in the balance. Heaven had spared us, and God's rain stayed with us the whole night.

The next morning I suggested to Leo that we go in his boat to Wittenberg to see if the village was still standing, and above all to help, if that was possible. Wittenberg still stood. The fire had only reached north a stretch of about five miles between us and the town. As we rounded Tower Rock, however, a tragic view met us. Only the masonry chimney remained of the doctor's little house, our first home. The rest was a smoking pile of ruin—a memento mori.

Harvest, Snakes, Pigs

German readers must surely imagine that we would have lived through years of a fire-wrought desert, but in an almost subtropical climate and at the right time of the year, when the sap rose throughout the plant world, things happened differently. After only fourteen days, the burnt-over ground in the woods was covered in green again. The shrubs, stripped of leaves, were sprouting anew, and a few steps from the road there was little of the fire damage to see, except for the experienced. Remaining marks of the fire were only the ruins of the eleven farmhouses on Cinque Hommes Creek, the valley south of ours, and that of the doctor's little house, and, of course, the permanent damage suffered by the surviving trees that had been deeply struck by the fire's force. The big fire south of us had doubtlessly been caused by track workers, and the one to the north by the old, half-blind negro, Dick Wilson, who had wanted to burn away the undergrowth around his cabin.

A peculiar and truly likable characteristic of the people of Missouri in this area, however, is that such things almost never have legal repercussions. Even though a good dozen farmers had suffered damages of thousands of dollars, to my knowledge the Frisco line heard not a word of complaint. "It's hard to prove that track workers started a fire," said the farmers, "and the railroad, with its first-class lawyers in house, has a lot of leverage." Only Dick Wilson, a former Pullman conductor, caught the anger of the people of Wittenberg, since his carelessness had seriously endangered the whole village. It was implied to him that he had better not let himself be seen in town, and the storekeeper, Mr. Müller, shut off his credit. The old man was carried over to Grand Tower in a boat three days later by a pipe-smoking daughter-in-law.

In June and July our main work was cultivating. Corn is surely one of the fastest-growing plants known, but the weeds grow even faster. To give German readers any sort of idea of their mass and toughness is impossible. Lumpweed and pigweed were the greatest sinners. I do not know their botanical names. There is also datura, a bushlike, poisonous plant that grows to heights up to eight feet, and there are vines that twist around the young cornstalks and bend them over.

Cultivating with the tractor requires a fine touch. The machine doesn't travel along beside the rows; rather, it takes the rows between its wheels. This is the reason for the high ground clearance of at least two and a half feet of American tractors. The cultivator blades, on each side, must be adjusted very exactly, so that on one hand they cut as deeply as necessary beside the corn, and on the other hand they don't damage the roots. The extraordinary value American farmers place on their string-straight rows is explained by the fact that good cultivating can't be accomplished if you have to drive wavy or zigzag lines. It pained me in the beginning to see how many corn plants I tore out of the ground while cultivating. It took weeks before I developed the necessary feel for driving, and, typically, Huc acquired it before I did. Machines and their operation are simply in the blood of young Americans. The boy vastly preferred the tractor over the car; I could always trust that the machine would be greased, the oil in its air filter replaced, all its screws tightened, and all its parts adjusted to a hair. More and more, Huc's specialty was cultivating, but that didn't mean that Rita and I were out of work.

Tractor cultivating dealt only with the space between the rows. Within the corn rows themselves the weeds proliferated just as before, and there was nothing else to do but to pull them by hand. For weeks, Rita and I stooped in the furrows, now with the right hand, now with the left when the right hand was tired, and still it didn't always get the job done. Our hands were constantly brown and green from the sap, and no soap helped. And when we observed our day's work in the evenings, barely able to straighten our backs, we were shocked by how little we had done. I might give an idea of the potency of the weeds on our land when I say that the edges of the fields had to be cut with a scythe every two weeks and that the piles reached my knees each time. In all, we cultivated and pulled weeds three times in this year. A good American farmer cultivates five times a season; we simply didn't have the strength for that.[1]

1. The work these three people attempted, not to say accomplished, was astounding. Huc Hauser was also only thirteen years old in 1946. Let us say

Our corn was three feet high at the beginning of July, and by the end of the month it was the height of a man and stood in full bloom. In the best field, near the river, the stalks were as thick as my wrist and grew to nine and ten feet. The sunflowers that I had planted at the edges of the fields hardly reached that height, but their buds opened and broadened to giant platelike flowers. Our dark green fields with their golden borders offered a magical view. Even the meadows, through repeated mowing, were improved and preserved. The old, cultured plants like clover won the upper hand once more over the weeds that had choked them for years. This showed how rich the land had originally been. We could have kept several cows, but we didn't have cow stalls or the necessary capital. There was also a written law that farms that lie alongside railroads have to be fenced suitably to contain large animals. With the size of our farm, this would have cost thousands of dollars in barbed wire alone.

In July and August, when no more cultivating was possible, the field work allowed a noticeable recess. We used the break to build an equipment shed; a long-term plan for which we had just never had time. In general, American farmers leave their equipment outside, year in and year out. It always bothered me, though, to see the hay rake or the mowing machine standing in a field in the middle of the winter, rusting in the snow. Huc also insisted that our loyal tractor had earned a roof over it. We began with the leveling of the ground, and we spared ourselves much labor by towing the cultivator over the higher areas to loosen the soil, then using the "scoop" to shove the dirt into the low places. The "scoop" is a big sheet-metal shovel that can hold about a cubic yard of earth; it is chained behind the tractor for towing. Next, we skidded the stones for the foundation to the site. There was no lack of limestone in the various creekbeds. It only cost sweat to wrestle the hundred-pound blocks onto the sledge, and a great deal of driving skill to get the load safely home. The last flood had floated a good dozen wooden beams onto our land, and while the water was still high I had dragged them up to the pasture so they wouldn't be stolen. These forty-foot timbers matched the

---

then that there were really only two and a half adults working, albeit the "half-man" apparently did the work of two.

dimensions of our shed, and since they were easily a foot and a half wide and a foot thick, we needed the jack to raise them to the level of the stone foundation and bed them in cement.

For the vertical structural beams, we once again used railroad ties—the longer sort that are used for switches—and with these we gained a roomy height of some ten feet. After unending sawing, we had prepared the necessary number of diagonal support timbers and braced the structure with them as was needed. Now, however, came the greatest difficulty: the lifting of the heavy girders that were to form the tops of the walls.

After we had tried every sort of supporting device and blocks and tackle and had almost been killed by their slipping and sliding, nothing remained but to saw diagonal scarfs and reduce the length and weight of the timbers by half. The roof trusses we framed on the ground and raised in one piece. However, we had no material with which to cover the roof. All the boards left as slash by the sawmills were used as sidewalls. In this deficit, I thought about the burnt farmhouses in the Cinque Hommes Valley.[2] I remembered that some of them were roofed with corrugated sheet metal and that it could not have burned. These farms had been repossessed by the state many years ago, like all farms on which, for five continuous years, the modest taxes couldn't be raised. Therefore, this old corrugated roofing belonged to Uncle Sam, if it belonged to anybody, and I had the feeling that the Uncle didn't place great value on it.[3] The old road over the hill had long since fallen into disrepair, and Huc and I often had to drag away uprooted or fire-toppled trees with the tractor, until we finally reached the deserted burn sites. Huc leaped grinning onto the heap of rusted steel sheets, but then he suddenly stood stock still. Between the top two sheets on the pile lay a coiled rattlesnake, which had duly warned the boy with its spooky buzzing. Following Huc's pointing finger, I aimed the shotgun, and as the shot hailed against

2. Hauser translated this French name into German, calling the valley of Cinque Hommes Creek "Fünf-Männer-Tal," or "Five-Men Valley." The farm is nowhere near the creek or valley of that name, however; the creek is several miles to the north.

3. Hauser stated that these farms had been repossessed by the *Staat*. He was in error in this regard; the farms would have been repossessed by the *Kreis*, Perry County, and not the *Staat*, the United States.

the metal, there was one less "rattler" in the world. It was the biggest rattlesnake that we had killed until now. It was thirteen years old. It was a good five feet long and was as thick as Huc's arm.[4]

Carefully at first, we lifted the sheets with sticks, but finding no more rattlesnakes we had undisturbed luck. The sheets not only were badly rusted, but also were totally warped by the heat. Nevertheless, we gathered together a considerable load and hung it on the tractor with chains. Huc, who is a born junk collector, also gathered dozens of door hinges, angle iron, and other sorts of scrap metal, and with endless care the jangling load was brought home. To me it was doubtful that the old sheet metal was worth the trouble of removing the rust, but the boy discovered a work-saving method that I can only recommend. In our creekbeds there were large deposits of sharp-grained sand. Huck fastened a dozen sheets together with wire and then dragged the sheet metal train in circles until they were sanded bright. Three expeditions to the burnt farmsteads brought together enough sheets for our roof, and a few days later it was radiant under a coat of primer, decorated with its christening wreath.[5]

Out of the remaining bridge timbers we built a "Repair Workshop for Automobiles and Tractors," as Huc proudly called it, at the back of our equipment shed. It consisted quite simply of two broad beams that were sunk horizontally into the hillside at one of their ends, while the other ends were supported by ties. The beams were laid at the width of a car's wheel track, so that in this manner we had built a repair pit. We could stand quite comfortably, and almost upright, to work under any vehicle we drove onto the beams. The affair proved itself so satisfactory that even distant neighbors brought their vehicles to our "Workshop" when there was something that needed repair on the underside of the car. Typical of all of our construction, we men did all the heavy work, while Rita did all the measuring and marked all the timbers for sawing. She possessed an inborn talent

4. Hauser apparently had assumed that the snake was thirteen years old because it had thirteen rattles. Indeed, a new segment is added at each shed, but rattlesnakes may shed up to five times in a single year, and buttons might break off as well.

5. It is a German tradition to dedicate new houses and other buildings with a wreath of evergreen boughs (a *Baukranz*) or a small pine or fir tree.

Huc, hanging the *Baukranz* on the newly built garage.
Photo courtesy Huc Hauser.

for such work and it carried over as well for furniture joinery as for house carpentry.

Huc's biggest summer project, though, was the boat. Since he had come home to us with his eleven-dollar and one-and-a-half horsepower gasoline engine, he had talked day and night about the boat he wanted to build all alone and with no adult help. And since he had worked splendidly for six weeks, I gladly let him have his way. Poplar wood, of which we had plenty from a huge felled trunk, would naturally be the obvious choice. My son, however, wanted his boat to last for eternity, and so it had to be cedar. The felling of oaks had severely damaged many cedars, and so we had enough of them. Cutting the trees was thus not a problem, although getting them home was. They were so unfavorably placed that we couldn't get to them with the tractor. The three of us almost broke under the weight before we had enough of the necessary trunks onto the road. Then another problem emerged: There were no more sawmills nearby. We could easily drag the trunks to the Altenburg sawmill behind the tractor, but not the finished boards, and these were too long for Perfidio's roof. So, building a farm wagon emerged as the more urgent first task, especially since the harvest was drawing near. Huc and I therefore went quickly and often to farm auctions. Modern farm wagons, with steel frames and roller bearings, cost far too much money, but we finally found an old model with a wooden frame and plain bearings for twenty-five dollars. It lacked a body, however, and this resulted in more work than we had foreseen. First, we removed the tongue, intended for horses, and put in its place a natural fork from a cedar tree. Huc "found" the necessary hardware for attaching it to the tractor, as usual, at the sites of the burnt-down farms. The hardwood boards for the wagon box we gathered as well, but when the box was finished, it turned out to be too wide, so that on turning, the front wheels hit it and gave the span too wide a turning radius. The whole thing would have to be built all over again—but before that could happen, the harvest overtook us.

The dark green of the corn had changed by August to a faded green. The stalks dried out and turned yellow, and where in July the purple tufts of the blooms had waved in the wind as proud flags, now the heavy ears drooped downward. Our fields didn't look so clean and orderly as those of the Altenburg farmers', but when the Webers and the Hoehnes came to our valley for fishing one Sunday, the men respectfully nodded: "Not bad. Not bad at all for the first time." And I was proud of those words, prouder than I have ever been of any book.

We began the harvest earlier than was usual in the region, and for a specific reason: One evening Rita forgot to gather the eggs, and sent the boy to the henhouse after dark. He came back with a look of disgust in his face. "There must have been a dead fish in one of the nests. It felt really smooth and slimy in there and it stank. I couldn't see it."

A fish in the henhouse seemed quite unlikely to me. Gripped by an uneasy feeling, I took the flashlight and shone it around the henhouse. In one of the nests lay a big coiled copperhead with flashing green eyes, the most dangerous sort of snake we had. Our son had literally held this thing in his hand and had remained unharmed. When the snake had been killed by a gunshot and the flapping chickens had calmed themselves once again, I stated, still somewhat pale, "Well, I've had a bellyful of this snake business. It must be because of the fire that they have all come to us. Now we're going to let the pigs loose. There are no better snake exterminators."[6]

Once said, so done. The next morning, our herd of swine, now multiplied to twenty-four head, were let loose to roam alongside the snakes in an altogether thankful mood. Swine are probably not immune to snakebite, but the bites can barely penetrate the fat layers, and the venom can't reach their veins. In Germany you could never afford to let a herd of pigs loose in a field. Here, though, there was so much natural feed that the animals caused relatively little damage. The blessings of nature were much richer than we could ever harvest. The pigs kept exclusively to the fruit trees, mostly apple and pear trees, that stood everywhere in the woods and which now were dropping their fruit. What in Germany would be an unthinkable and revolting sight, pigs in great bunches smacking away on fallen peaches, was perfectly normal to us. By the time we let the pigs go, Rita had already canned up to five hundred quarts of jam and marmalade. There were no jars for more, and for shipments to Germany, fruit didn't really even pay for the freight.

After a while, however, the pigs began to go into the cornfields for the watermelons and for the corn itself, but only after they had been led there by the dogs. As unbelievable as it sounds, bloodhounds love corn, and they have a highly developed technique for tearing down cornstalks and husking the ears. From this time forward, there developed a strange symbiosis on our farm, in which none of the animals had to be fed: Kitty and Holla tore down more corn than they could eat. The pigs ate what the dogs left behind. The young pigs left behind corn grains as much as if undigested, and the chickens pecked their feed from the swine manure. The economy was complete. The corn had yet another devotee, the raccoons. Raccoons search for

6. Huc Hauser recalls this incident quite well, but somewhat differently. First, he had been sent to the henhouse after dark as penance. It was his job, not Rita's, to gather the eggs in the afternoon. Second, his comments to Rita and his father were made because he thought they had played a joke on him with the "dead fish" as part of his punishment. Third, the snake was a rattler, not a copperhead, and it slithered away as soon as the light appeared. He has a vivid recollection of watching the reptile's rattles silently disappear over the edge of the nest.

food almost exclusively at night, and thanks to their extraordinary cleanliness, they do not cause too much damage. The racoon climbs up the cornstalk until it reaches the ear and gobbles it without losing a single grain.

After consideration of our successful swine production, and also of our limited powers of labor, we began the harvest in late August, while an American farmer would have waited gladly for the first frost, when the corn would have been fully hard and dry. Huc drove the tractor and the wagon down the length of every second row in the field, while Rita and I broke off the ears on the rows to the left and right, and Huc took over the downed center row. Corn picking in the United States has long been considered a contest. There are annual corn-picking tournaments, especially in the west. Infected by this tradition, we also treated the work as a competition. The work is in no way easy when the slow-moving tractor smacks you in the face with a hard stalk, or a razor-sharp leaf slices into your arm. Of course, we didn't harvest the whole day. At noon our farm wagon was full. It held about thirty bushels, and Huc would then have to drive it into Altenburg to Fischer, the grain dealer. That was a boring business, but also not without danger. The shortest distance, ten miles over the southerly track up the wooded hills, in many places had grades and descents of over 20 percent. Since the farm wagon had no brakes, on deep descents it would begin to push the little tractor sideways and try to overturn it. This required considerable driving skill, and toward evening I became so nervous that I would head a couple of miles in the boy's direction, until I could hear the putting of the little machine and finally exhale. Huc would already be waving happily from far off with the purse he had dangling from a strap around his neck that bore the day's receipts. Mr. Fischer was the picture of understanding, as he was willing to sack our ears and weigh them individually, instead of weighing the wagon on a truck scale and comparing the difference between loaded and empty weights. In this Mr. Fischer was unquestionably honest, and he paid in cash. The price of corn around this time came to some $1.60 per bushel, but since our corn was not quite dry, Mr. Fischer held back twenty cents for weight loss. That was disappointing.

We had less luck with our garden produce in Altenburg. The whole area around there had a vegetable surplus, and even though buyers from the St. Louis Farmers' Market came twice a week, our beautiful tomatoes brought barely a dollar a bushel—barely six cents a pound. Once in a while Leo Harris took a boatload of cucumbers to Grand Tower, but while cucumbers in June still cost five cents apiece, now a whole bushel could be knocked down for half a dollar. We couldn't even think about fruit sales, because our trees had not been sprayed against pests, and their fruit was therefore imperfect. So, we dealt with it all using the good old American maxim: "We eat what we can, and what we can't, we can." For hours, Rita squashed hundredweights of tomatoes through a wire mesh stretched over a washtub, looking as if she were sprayed with blood from top to bottom. Tomato paste was at least a con-centrated, weight-saving form of preserving, so that sending it to Europe could be profitable. For hours in the evenings we turned the wheels of the sealing machine for the cans. Interchangeably the kitchen smelled of raspberries, blackberries, peaches, or beans with bacon, each according to what had just been preserved. In order to rescue my wife from the hot kitchen at least on Sundays, I played master chef in my own style, as I had learned to when I was in Chile.

Preparations began on Saturday afternoon or Sunday morning with Huc and I together shooting a good take of game. Partridges were especially liked for this purpose, but also wild doves or a snapping turtle. Then, about ten o'clock in the morning, we would light a substantial fire in a depression dug in the gravel bed of our creek. Next, we gathered the best that garden and woods could deliver—the juiciest young corn, the freshest vegetables, wild grapes, and even hazelnuts and walnuts. When all this was cut up and washed, and the birds cleaned and plucked, we took Rita's biggest cooking pot, threw in a big glob of butter, and laid our treasures inside on grape leaves until the pot was three-quarters full. Salt, pepper, and some seasoning herbs were added, and finally a bottle of cheap California red wine poured over it. A bottle cost twenty-five cents. Meantime, Huc had mixed a handful of flour and water to a paste. The lid was placed on the pot and sealed airtight from outside with the flour paste. The last thing was to fasten two wire hangers to the pot handles—as will soon be seen, this was very important—and the vessel carried to the dry creek bed.

The fire, meanwhile, had burnt itself fairly well down, part of the glowing ashes had fallen into the cook hole, and all the rocks round about were glowing hot. Carefully we lowered the pot into the hole, and with our eyes pinched shut, shoveled glowing ashes and gravel over the pot until it was completely covered and only the wire slings could be seen sticking out. Keeping our eyes tightly shut was necessary because the slightest touch of the shovel caused the glowing gravel to explode like shrapnel, and many splinters would go swishing past our ears. Then we brought the dishes from the house, prepared a comfortable seat for Rita in the shade of a cottonwood, and waited with excitement and impatience for the great moment, an hour and a half later, when the shovel handle was passed through the wire slings and we used our combined strength to haul the treasure out of the smoking ground.

Even if I, as the cook, say it, a stew like this is poetry. It must lie in the complete air seal and the great heat from all sides. In any case, the vegetables on top were covered with a sort of a sugar sauce of their own juices and the wine. Best of all was the thick brown sauce on the bottom of the pot, which in the American South is called "pot likker," pot-lick, a special treat.[7] The primary error in my kitchen was that Huc and I always cooked much too much, so that we had to live on such a pot for days.

On one such Sunday afternoon, it happened that we heard one of our dogs, who were rummaging about in the area as usual, cry out abruptly in a howl of pain. A couple of minutes later, Holla, Rita's dog, came running up whimpering and crawled to Rita's feet with a large lump on the back of her head, which we could see swelling. Beside himself with anger, Huc jumped up and went to get his rifle. "The snakes have to die!" "Stay here," I ordered. "Do you want to be bitten, too?" We carried the whimpering dog to the house at a sprint, and there we tried to lay the wound bare. With the help of a razor blade I found the place behind the ear where the typical triangular bite had penetrated. With the help of a heated egg cup

7. Hauser's translation of *pot likker* was *Topfschlecke*, "pot-lick"; he believed the term to mean that something was licked from the pot. Ethnologists and linguists, however, do not agree over whether this expression originally referred to pot *licker*, or pot *liquor*.

we were also able to draw out the wound somewhat, but it was already far too late. Within an hour, the poor beast swelled to nearly double its size. With its eyes shut it lay there, and its tormented heart pumped ever harder. According to hunters' lore in our region, you have to fire a shotgun as close to the wound as possible, so that it will be cleaned and burnt out. In our first shock we didn't think about this. Also, I hold this tactic to be a superstition. Streaming with tears, Rita did everything she could imagine to strengthen the failing heart. Now it was strong coffee, now whiskey that we tried to get into the animal, but its swallowing muscles had already cramped closed. Night fell, and Holla still lived. She lay on a blanket next to Rita's bed, and again and again during this night I saw my wife kneeling beside the bed with a flashlight. Against all expectations, Holla was still alive in the morning. Coverings of a clay and vinegar paste seemed to relieve her somewhat. Leo came around noon, contemplated the animal, and said, "If she's lived till now she'll probably come through it."

For three days Holla lay rigid, and only when she was in a condition to lap up a saucer of milk did Rita lie down for a first real sleep. After that the bloodhound quickly recovered superficially, but it was no longer the old Holla who began to play with her sister Kitty on the fifth day. A part of the inner light had faded from her sad, beautiful eyes. The animal remained as amicable as ever, even more affectionate than ever, but it had lost in intelligence and in tracking ability. And just as a mother loves her crippled child more than her healthy one, Rita was even closer to Holla than before. It couldn't be denied, though: Holla was useless for hunting and a little bit dull-witted.

Huc had used his free time in September mostly for the building of his boat. He had provided himself with construction plans from his favorite magazine, *Popular Mechanics,* but discarded them as being overelaborate and unworthy of his own, superior boatbuilder's art. As the boat grew, it took on an adventuresome, and, from a seaman's perspective, a never-before-seen form, so that I sometimes had trouble hiding my head-shaking. Obviously it would not be completed before the end of the summer vacation, and secretly I thanked the Creator for that. With its eddies and constantly changing sandbars, the Mississippi is not the Rhein; even the lower Elbe is harmless in comparison. Huc's wild determination to go on a big trip with this boat would, I hoped, be somewhat dampened by next year.

The departure was sad for us as usual, and as we came home from the station and stood in the loneliness of our empty fields, Rita said, "Now it's fall, and I think we should give PeeWee his freedom. He has a right to live a natural rabbit's life."

PeeWee had been our housemate all summer; one might well say, our house tyrant. At each meal he sat on the table like a decorative centerpiece and took his tribute from every plate. He had learned as well to leave his little pellets in a special corner on newspaper left for the purpose. PeeWee was the most human rabbit that ever lived, and he showed it in a manner that brought tears to the eyes of Rita, who was already sorrowful in these days. As she carried him to the garden as her greatest offering to his right to freedom, he ducked down into a furrow, and far from enjoying his liberty, there he sat completely motionless with an anxiously pumping heart. That evening he lay in exactly the same spot. He had not so much as wiggled his nose at the cabbages. PeeWee only recognized vegetables served on a plate. As it became dark, Rita carried him back into the house, on the grounds that "PeeWee is afraid of the night."

This went on for three days; every morning PeeWee was taken to the garden and lay fearfully crouched in the dirt until evening without moving. On the third day, something strange, indeed, almost magical, happened. By incident we were able to observe this out of the kitchen window. The afternoon had brought the first cold fall wind. The plants in the garden shook themselves and bent over in the half-storm, and PeeWee's fur was ruffled by the wind gusts. Suddenly the little fellow raised his head and sniffed in the wind. In the next second, he tucked his ears back, leaped into the air with all fours, raced off on the double as if the Devil were after him, and disappeared into the woods through the nearest hole in the fence. We never saw him again. After that, however, I never shot a rabbit within a mile's distance, out of fear that it might be PeeWee. Rita never wanted to eat rabbit again, either.

At the end of November came the big day of the hog auction at the county seat of Perryville. I had reserved a truck for this because my farm wagon had become too small for the herd, and because I needed another good man, anyway, for rounding them up and loading them.

We loaded seventeen hogs, the heaviest carrying three and a half hundredweight, and the lightest weighing at least two hundred fifty pounds. Swine prices that fall lay around twenty-five cents a pound, live weight. I only really understood about half of what actually happened at the auction, because it all happened lightning fast. As the cashier pressed into my hand the sum of $1,257 in bank checks and bills, it was as if I were dreaming. The swine had cost us as good as no feed, had required as good as no labor, and my original capital had only been a hundred dollars. These pigs were by far the most successful speculation I ever made in my life. I couldn't get to the bank fast enough, and I was enormously proud as the bank manager tore up my mortgage before my eyes and threw it into the fire.

Since in the meantime all the fields had been harvested and all but the reserved seed corn sold, it was possible to draw up our balances for the year. We had sold twenty-one hundred bushels of corn at an average price of $1.50 a bushel. Together with the proceeds from the swine, we had taken in $4,407, not mentioning the vegetables we had sold. Now, admittedly, we didn't have an orderly bookkeeping system and had calculated neither tractor fuel nor our labor, but the result was in no way bad for inexperienced farmers in their first year. The packets to Germany could proceed. We now provided a large number of families with monthly shipments of twelve to fifteen pounds, mostly fats. Since the value of these shipments was tax deductible, we only had to pay tax on $2,500. If I remember correctly, we paid altogether $143 in income tax, and $28 tax on the farm. It was a good, a blessed, year.[8]

---

8. In the last decades of his life, Hauser had turned more and more toward a sort of Jeffersonian belief in the cultural and social value of a free agrarian society. His land in New York had been too little, too cold, too rocky, and too uncomfortable, but this milder and joyously fruitful land in Missouri renewed his faith in that ideal. After his return to Germany, he insisted that he wanted to go back to the farm and live and work there for the rest of his life. He never returned, however, and when Rita deeded the property to eighteen-year-old Huc, the son discovered that it was mortgaged, and he had to sell it to clear the debt. As Hauser's biographer Helen Adolf stated, the farm was another of many changing mirages in the man's life.

Roots in the Ground

Mail from Europe came fairly regularly in this second post-war winter. There were at any given time whole packets of letters, and after they came it was usually the case that Rita and I, each alone within ourselves, rushed off into the woods or to the creek in order to work through the turmoil of our feelings. Only in the evening, by lamplight, as we sat near the stove, could we bring ourselves to exchange our letters and discuss the news. It is not necessary to state here that this was about all the miseries of mankind, so unfathomable and so unquenchable, that everything we tried to do to help seemed as if nothing to us. We were ashamed of our own security and contentment. We often remarked that it would have been better if God had let us be a part of the destruction of the Old World.

At my rough-planed desk in these months I wrote in a race against necessity, and only for the money. A good book, a work of art, would have wasted far too much time and brought in far too little. I wrote wild adventure stories and science fiction novels at that time. Every six weeks I produced a serialized novel for cheap, sensation-hungry magazines like *Amazing Stories, Astounding Stories,* and whatever else they may have been called. In addition, I produced translations for a Chicago publisher, among them some that were rewarding and commendable: Max Picard's *Hitler in Our Selves (Hitler in uns selbst)* and his *Immortal Marriage (Unsterbliche Ehe),* Ernst Jünger's *Peace (Der Frieden),* and Friedrich Georg Jünger's *Perfection of Technology (Perfektion der Technik).* Nevertheless, I have to confess that even in these my theme every time was the money, a literary prostitution whose only defense lay perhaps in the fact that this money might wrestle a few stalks away from the grim reaper.

Occasionally we spoke with some bitterness about how easy it would be to feed a half-dozen starving German children on our farm; in Altenburg and Wittenberg and in the whole region there was barely a house that would not have taken in such a child. Big-hearted Americans had made hundreds and thousands of applications for admission of German children, but the "responsible agencies" had declined. The human order of the whole world was literally crazy, but even fundamentally healthy persons were brought to the edge of insanity by this crazy world.

The heavy depression that burdened us that winter, deepened by loneliness, only lifted when the meadows greened up again, when, with its bellowing engine, our little tractor pulled its first furrow, and when Rita came to me with a wonderful plan:

"I can make us another four hundred dollars a month on the side; think about that—it would mean another forty packages a month to Europe."

"How would you do that?"

"With Angora rabbits. I sent for the brochure from the breeders' association. They supply the breeding stock for fifteen dollars apiece and guarantee the purchase of new litters and wool at daily prices. And the best Angora wool brings eighteen to twenty dollars a pound today."

"Do you think you can do all that along with all the rest of the work?"

"Absolutely. Here it says, 'practiced hands can shear an Angora rabbit in only three minutes.' What they can do, I can do. The only thing we need at first is a big open shed for two to four hundred rabbits, maybe. If we start with six breeders, we'll have that many by fall. I've already figured out all of that."

For the shed posts, I used young cedar trunks, which won't rot in the ground. We covered the roof with side slash from the sawmill. The shed, which stood at the bottom of Cemetery Hill, was a good twenty feet long. Roofing paper was the only cash outlay for it. The wire mesh for the cages, which Rita built from the plans from the breeders' club, was expensive, as were the copyrighted plans themselves. By the time the first animals arrived by express, six does and a buck, almost a hundred dollars had disappeared from our narrow bank account.

But when the ruby-eyed, satin-soft creatures arrived, half dead from the trip and shivering from excitement, it was as if the household suddenly had acquired as many new babies. From then on, Rita was to be found either in the Angora shed, or, rather, armed with a sickle and a basket, looking for feed for her darlings. The mother does had already been covered, and each of their transport cages had a little note with the expected date of delivery. Even though the best midwife in the world couldn't have helped the mother bunnies with that, Rita set up a cot in the rabbit shed. As the first mothers-to-be began to upholster their nests, and as the first litters were well on their way, we took them into the house "to play." There our dogs, with soft mouths and mildly reproving eyes, brought them back into view when they crawled under the furniture.

In terms of human relations, the Angoras were a gigantic success. In financial terms, however, they were unfortunately something altogether different from what it said in the prospect. The highly bred animals refused fresh fodder almost entirely. The oats they needed didn't flourish in the warm climate of our farm, and factory-made pellets of fine-ground clover cost ten times as much as fresh clover, of which we had a surplus. When it came to shearing, each bunny was placed on a sort of barber's table and had to be petted to calm it while Rita sheared. The procedure lasted for hours, and often made my wife break into tears when a skin injury occurred despite all her care. A farmer's wife really ought to be made of harder stuff, but to Rita's credit, our Angoras cried out in fear and pain like children, in a manner that was really heartrending. By the end of May, Rita had gathered a pound of Angora wool for shipment, carefully sorted in paper packets in first, second, and felted quality. Eight days later came the devoutly awaited letter from the breeders' club, and it contained something like this:

Valued Member,

We thank you for your first shipment. We regret to inform you that the price for Angora wool has fallen by 50 percent in the last month. You have sent us 6 ounces of wool of the best quality, and 6 ounces of second quality as well. We can no longer accept wool of lower quality. The value of your wool at today's price amounts to 3 dollars and 86 cents. From this we subtract our administrative cost of 80 cents. Enclosed is our check for 3 dollars and 6 cents. Thinking always of the concerns of our members, we close with great respect.

While Rita stared stunned at the paper, I considered the contrasting side of the account: feed costs to this point, twelve dollars;

construction costs, around two hundred dollars. That the Angoras were a losing proposition and that it would be best to get rid of them as quickly as possible was absolutely clear to me, but at that moment I didn't have the heart to say it to Rita.

The spring planting, fifteen acres more than last year, had already ended when our son arrived for the summer. As usual, the chain of "great events" for him began at once.

We had a picnic on our highest riverbank that first evening. Later, satisfied, each with a canteen of cold, sweet tea, we lay on the grass almost three hundred feet above the river and watched as the vast ball of the sun sank behind the Illinois hills, as the Mississippi turned to molten silver, and as the wild geese spread themselves to sleep on the sandbanks.[1] As his latest procurement, Huc had brought a telescope, which he swept upstream and downstream, and suddenly he shouted:

"Here comes one of the old stern-wheelers. It has a long string of lights. It must be a passenger steamer."

The old stern-wheel steamers, of which Dickens and Twain had written, are today as good as extinct. Many years ago, I made an unforgettable voyage on one of the last ones, the *Tennessee Belle*, from New Orleans to St. Louis. At this time, to my knowledge, there was only one left, the *Golden Eagle*, and the newspapers had written that this year it was going to conduct pleasure excursions into the Old South. I took the glass: crowned double smokestacks, a Chinese pagoda for a pilothouse, a two-story superstructure, and a paddle wheel as high as the ship itself. Indeed, this could only be *der Goldener Adler*.

Fifteen minutes later, in the quickly descending darkness, it came blustering by us, its hundred-year-old steam engines gasping heavily. Garlands of colorful lights hung along its decks, completely

---

1. Immediately outside the farm's gate, the road north to Wittenberg climbs sharply and winds through the thick woods and undergrowth. To this day along this stretch there is a pullout where motorists have left the main road for years. About thirty feet to the east is a crest with a break in the trees and a magnificent view of the river, the railroad below, and the Illinois shore more than a mile distant. The view from that spot, as Theodore Roosevelt once said, "bankrupts the English language." My heart tells me that this must have been the picnic spot.

cloaked in the veil of spraying masses of water. It was at least two miles away, holding to the deep, navigable water on the Illinois side, making its way through the ships' channel between the shore and Treasure Island.[2] Even so, through the glass, it was as if we could see the passengers dancing.

After a few minutes the radiant vision disappeared behind the island, its wake rolling over the sandbanks on our side with a dull rumble. "I wish the three of us were on board," Huc mused.

At that very moment it happened. A crash and blast behind the curtain of the woods on the island; a powerful hissing release of steam; obscure, shrill cries; fearful, cursing men's voices; and the flash of a spotlight that was just as quickly doused. Then there was stillness, total and eerie, the hoot of an owl the only sound. We watched the south tip of the island tensely. The current in the channel, we knew, ran a good eight miles an hour. The *Golden Eagle*'s own speed was at minimum twelve miles an hour, and the island was three miles long. The ship must appear in just a few minutes . . .

A quarter of an hour passed, then a half. Nothing showed itself. Meanwhile, it had grown fully dark, and off to the south rose a wall of clouds with immense sheet lightning. "I'll bet it was a boiler explosion," said Rita, who always seems to think of the worst.

"I believe they just ran aground," I said.

"We have to go over there right away, Father," Huc demanded. "We have my boat; maybe we can rescue people."

"Your boat hasn't ever even been in the water. We have no oars or anything. It will probably sink right away, and even if it floats, we'll be much too late to save anybody. Besides—you can see the channel from Grand Tower. If anything did happen, the fishermen are surely already under way."

Even Huc had to agree with this, especially since just then the sky opened its sluices with an authority that one only sees in the tropics—and on the Mississippi.

The next morning I was awakened before dew and day by dogs barking and by a deep bass voice.

"Hey, there, Hey there! Is anybody home? Come on out, Boss, I got something to talk to you about."

2. This is the same island Leo Harris called "Pig Island."

It was Black Bill, every inch the pirate, a whiskey bottle in his hand and stumbling in blissful drunkenness.

"Ha, Boss, there you are, finally. Man, you're sleeping away all your good luck! Take a drink, and when the bottle's empty, throw it away. There's a lot, lot more where that came from. What do you want, Boss, perfect beds, first-class spring mattresses? What's the Lady Boss want? Cooking pots, copper kettles, jam, knives, forks? Everything as good as new. But dirt cheap—almost a gift. My boat is full of the stuff."

"Bill," I said firmly, "did these things come from the *Golden Eagle*?"

"So, you know already? Probably heard it last night? Ran aground, it did, broke its back. The *Golden Eagle* won't ever float again."

"How did it happen?"

"Rudder chain broke, right in the narrowest part of the pass, the pilot says. Others say the pilot was drunk. Could also be a put-up job, on account of the insurance. We'll probably never know."

"Were any of the people hurt?

"Ha! Not a trace. There were 120 passengers on board, and after they grounded, they all just walked off. The mosquitoes caused the ladies a little bother, and they griped like fishwives. A bunch of boats from Grand Tower got there last night and took the crew and passengers away."

"Yes, and then?"

"Yeah. Well, Boss, then the wreck was left alone, and naturally we cleaned it out a little bit. It would be a shame if at the next high water the whole thing went under. Ain't that right? What you say? You really don't want nothing from all that good stuff? Aw, that's dumb—okay. I'll be on my way—otherwise the others over there will drink up all the whiskey."

Naturally, what I feared took place. In all the noise, the boy awoke and heard the whole story. Then there was no more holding back.

"Father, we have to go see it. We have to go to the island, with my boat."

While Rita quickly fed all the animals and made a big bundle of buttered bread for us, Huc and I slapped together a pair of oars in all haste. They were nothing but a couple of pieces of board nailed to young cedar trunks. At the last minute, it occurred to me to take along the shotgun and a few shells. Then we loaded the clumsy boat onto our wagon, dragged it with the tractor as close as possible to the inlet under the railroad bridge, and, before we had given it much consideration, we had already put in. Rita was at the rudder, and Huc and I were at the oars.

We were already a couple of hundred yards out and being driven startlingly fast downstream when I noticed how heavily we were taking on water. Dozens of little springs shot up from the seams, and some of the bottom boards appeared to be loose, which after a year's weathering in sun and wind was no wonder. Even though their feet were in the water, Rita and Huc had been too excited to notice the danger. Fortunately there was an empty can in the boat, and since I simply had to row, I cried out to Rita: "Steer with one hand and bail with the other!"

With a suddenly startled face she did so, but the quart can couldn't keep up against the many leaks. The only right thing to do, naturally, was to row immediately back to shore with all our might. Meanwhile, though, a totally unrealistic ambition had taken me over. I had promised to get them over to the wreck. They trusted my seaman's skill and my boy was as proud as a king about his boat. Lightning fast, I reckoned the distance to both shores. They were just about the same, and the set of the current was slightly toward the island.

"Keep rowing," I ordered, "as hard as possible."

The primitive oars bent to the point of breaking. Rita bailed for the love of life. The rucksack with our provisions floated between our feet. The shotgun had long been under water. Sweat streamed from all our pores. The deliverance of the south end of the island, which we absolutely had to reach in order not to be driven into the middle of the stream, lay nearly exactly at our side. We threatened to be driven right by it.

Then it happened. Huc had laid on too much force and exuberance in his rowing, the cedar oar ruptured at the nail holes, and suddenly he sat, stupefied, holding only the bladeless pole in his hand. At this moment, we were within a hundred yards of the shore. The water here swirled in eddies turned up by rocks below and ran especially hard. Driven abruptly off course, the vessel spun in a circle three times and threatened to capsize. I grabbed the remaining oar in breathless haste and worked my way to the bow to paddle on both sides. I had only made a couple of lightning-fast strokes to direct the bow towards land when suddenly I lost my footing. I hurtled right through the bottom and disappeared in the gush of water that at the same moment washed over the bow. Rita uttered a shriek. Emerging, I saw my love sitting in water up to her knees, pale with fear, saw the sandbar on the southerly tip of the island fly by, and knew that if we didn't reach it we were lost. Luckily, in this instance I had held on to the oar. I stuck it over the side, found the bottom, and in the next moment I was myself overboard. The bottom was slippery gravel, the water was nearly shoulder deep, and the clumsy trough of a boat nearly yanked my arms out of my body. Two or three times I felt myself twisted away, but I didn't let go of the boat and checked its movement as soon as I found footing again. I had thrown the oar back to Huc when I jumped overboard. By then the boy had so composed himself that he was able to help by poling, and now Rita jumped overboard as well. Inch by inch, combining our strength, we were able to drag the wreck onto the sand. Within feet of the chocolate brown eddies, we lay motionless and mute from exhaustion. Then our wet clothes began to steam in the blazing noonday heat, I rubbed my skinned shinbone, and, having lived through this horror, we looked at one another and broke out in irrepressible laughter.

"Never in my life will I forget the dumb look on Huc's face when all of a sudden there he sat with only the shaft of the oar in his hand."

"The Devil take that boat of yours, son! Do you know what really happened? A whole bottom plank simply fell out of your masterpiece."

"And Rita screeched as if she had been stuck on a spit."

"And rightly so," said Rita, indignantly. "I saw all three of us as drowned bodies. That was the first and last time I'll ever go out rowing with you men."

"We went looking for a wreck, and now we find *ourselves* ship-wrecked."

The boy hobbled over to his vessel. "Nothing to be done," he reported after awhile, "the bottom board floated away. But the rucksack is still there, and the shotgun, too. Father, what do we do now?"

"The only thing we can do is work our way around the beach to the wreck of the *Golden Eagle*. If we're lucky, the looters will still be at work and we can get one of them to take us back to the farm."

The rucksack was halfway watertight, and the buttered bread was still edible after all. Above all else, Rita had carefully stored a pack of cigarettes and a handful of matches in a rubber-lined inner pocket. As the food soothed our nervous stomachs and as the first wisps of smoke curled, our mere existence after such a menace seemed doubly wonderful. My soaked wristwatch had stopped, but the shrill whistle of the noon train over on our side told us that it was twelve-thirty and high time to get our trek underway if we expected to be back home by evening.

Three miles. That sounds like nothing, but on an overgrown island in the Mississippi such as this, one could well speak of "insurmountable obstacles." There were willow thickets right up to the water, so thick that we needed all our strength to force our way through. Then, house-high barriers of driftwood suddenly appeared, which we either had to go around or climb carefully over, and where we usually sank into the rotten timbers up to our hips. We would no sooner prevail over these than a stretch of quicksand came along, out of which we would have to scramble as quickly as possible. Then there were mud banks that the stream had washed up; the edges of these fell away under the weight of a footstep like an avalanche. Next, you would believe you had a few hundred yards of open ground, only to find, on drawing closer, that the opening was an inlet that forced you to make a half-mile detour, or worse yet, an old channel that crossed the whole width of the island. Such relict channels, out of which whole forests of dead trees towered, could be neither waded through, because you sank in the deepest, stickiest mud with each step, nor swum through, because you were caught up in the tangle of branches. Alternately bathed in sweat and muddy water, with all our clothing torn, with scraped skin, and bleeding from hundreds of horsefly and mosquito bites, it took us until sundown to make two miles. Then, in a particularly bad place, between two mountains of driftwood, Rita sank down onto a tree trunk, dangled her wounded feet in the Mississippi's water, and declared, "It makes no difference to me what you do; I'm not going any farther."

There was nothing to do. We knew Rita, and we knew she never gave up, except when there was really no way out. Even if I had been able to make the last mile to the wreck, the night would have long since overtaken me. Also, I couldn't assume that the beach bandits would still be at their work. Firelight or lantern light so close to the navigation channel would have been much too conspicuous. The only thing we could do was camp right there on that spot, and we had neither shelter nor provisions. The most important thing now was a smoke fire, not only to keep off the mosquitoes, which were unbearable in the evening and at night, but also as a signal, in case a fisherman might come by our way.

I called to Huc to haul up driftwood for the fire and willow branches for a camp shelter. Then I took the shotgun and worked a path inland, hoping that I could bring down a couple of squirrels for supper.

As I entered the forest, it was as if the night had already fallen. Giant oaks, cottonwoods, and hickories formed an unending dome with a roof like deep green bottle glass. If the island had not been so flat, I would have been able to walk much more easily in this forest than on the beach, but the annual flooding had fetched up barriers of driftwood around every tree trunk. With each step, dry branches cracked under my feet. There were swarms of squirrels, but since every tree was girded with vines and ivy, I saw the skittish red and gray shadows for only fractions of seconds. From their green palaces they mocked the clumsy, noisy human being, who had no better chance of bagging a squirrel than an elephant.

Troubled by the sight of the golden streak of the sinking sun that I might lose my way back, I pressed on. Unexpectedly an opening appeared before me, and I couldn't trust my eyes: At the far end of the clearing stood a house, shrouded all around by the woods. I knew that in earlier times, when the Mississippi floods were not so high and not so frequent, people had lived on the island, and that there had even been a schoolhouse there, yet even so, this unexpected encounter was utterly unearthly. It was a stilt-house, the floor built on posts the height of a man above the ground. This was probably the reason it had withstood the floods in the past. Its wood was weathered gray-white and covered over with the golden luster of the failing light. Loose boards clattered in the afternoon wind, and a part of the roof had slumped downward. Out of two windows, broken panes sparkled at me squinty-eyed in the dusk. Though at first glance I had inwardly rejoiced at the possibility of spending the night under a roof, now something sinister grasped me. I felt no doubt: A human-hating demon, perhaps the spirit of the wilderness itself, had seized possession of this house. I stood stock still, as if frozen to the ground. It seemed unthinkable that once this house had been filled with human warmth, that children had played in front of its stairway, that laundry had hung on lines, that in its clearing a plow had drawn its furrows. My footsteps were silent in the high grass of its former fields, and then the cry of an owl sounded from its gables. In the next moment, a frightened rabbit was driven out of a nearby thicket. After two or three hops it stood still, its white tail gleaming like a ball of silver. Instinctively, I swept up the shotgun and in its fire saw the animal catapult forward, even before the shot's echo from the haunted house sounded in my ear. That broke the spell. It was as if in killing its food the man had conquered the wilderness for a moment, and I suddenly perceived a trace of the feelings of those first deerskin-clad hunters who had opened the American wilderness. I hastily grabbed my kill and broke through the underbrush. In front of me was the last glimmer of daylight, and behind me the eerie night, falling in giant waves. It was like an exodus.

It was a big, quite well-nourished rabbit, and since it was impossible that it could be our PeeWee, even Rita partook of it. There was no salt, but just the same, no campfire roast had ever tasted so good to us. And never had we felt so strangely lost, and at the same time so secure, as when we crawled close together later in our shelter of willow twigs, protected by the smoke cloud of our campfire, and enveloped in the murmur of the river. Against our expectations, we slept well, nestled together like dormice. Our departure took place in the earliest dawn. There was nothing there for breakfast anyway, and pressing concern about our unfed animals at home drove us to get to the wreck and to find a boat home as quickly as possible. This time I struck a path over the relatively high central backbone of the island because less driftwood was heaped there. It was a fantastic passage. Millions of spiderwebs stretched between the trees and glistened with dew. We cleared our way through them with outspread hands, and it was good that the spiders were still stiff and unmoving in the cool of the morning, because these ugly gray tree spiders are really a bit poisonous. White herons burst in flocks from the watercourses we crossed, beavers swam hurriedly to their underwater houses, floating tree trunks came alive as hundreds of turtles, disturbed in their enjoyment of the morning sun, dove off in displeasure. The whole of it was a natural paradise, just as the earliest descriptions of travel in America portrayed it: raging life that induces reverence by dint of its primal power.

We reached the wreck near eight o'clock in the morning. The *Golden Eagle* lay spread out over the spit whose crown had broken her keel. The bow and foredeck were as if ironed flat, while the aftership with its giant stern wheel hung in the water. The crowned smokestacks lay like a pair of fallen dignitaries on their sides in the sand. Countless footprints, empty bottles, and shattered crates were witness to the activities of the beach pirates, but unfortunately there was not a single living person to be seen. Huc had naturally climbed onto the wreck first thing and was naturally just as compelled to go immediately aft, where his main interest, the old-fashioned steam engines, stuck halfway out of the water. Out of the corner of my eye I saw him open one of the heavy fire doors of the boiler. In the next second he recoiled and fell on his behind in shock. Out of the fire hole came crawling a gray-black apparition, with grinning white teeth the only bright spot. As it emerged out of the shadows into the sunlight, we quickly wanted to embrace it:

"Man, Leo, why are you crawling around in the firebox?"

Somewhat sheepishly, the old man first laid a crowbar and hammer carefully on the deck and mumbled, "I didn't know it was you. I heard voices and thought to myself that a police boat had tied up a little below."

"Is there still anything left for you to take, Leo? I thought the others would have pretty well cleaned it out."

Disdainfully, the old man spit tobacco juice overboard. "They're all dummies and boozers. They don't know anything about ships. They have no idea what's best on such a boat."

"And what is the best?"

Leo rolled his brown puppy-dog eyes enthusiastically. "Copper, naturally. Copper tubing and copper plate. I already have at least four hundred pounds, and if I sold it, that alone would bring sixteen cents a pound. But," he lowered his voice secretively, "I won't sell it at all. This copper will never leave the island."

"What will happen to it then, Leo?"

"Man, that will be my new still! And I've found a hiding place for it. The cops will never find it."[3]

Then we reported our adventure, and Leo shook his head, "I could have told you that before. Huc's 'boat'! I would never have said it while he was building it, but I wondered to myself, 'Why is he building such an almighty big hog trough there?' Well, I'll carry you over to your farm."

As the dogs ran to us yelping for joy, as the swine grunted, and as the geese greeted us quacking and the Angoras leaped up and down in their cages like white fur balls, we felt how our farm really needed us, how deeply we were rooted in our ground, and how the love it was given was returned a hundredfold. It was the happiest homecoming of our lives.

Huc's greatest catch this summer was called Chalk.

"Father, you won't believe it. I was sitting at my stand in the woods, and all around me it was so quiet I could hear the squirrels gnawing on last year's nuts. All of a sudden I felt a hand on my shoulder pushing me down, a blast of fire went over my head, and a squirrel that I hadn't even seen fell out of a tree. That's how quietly Chalk can sneak through the woods."

Chalk means chalk, and chalk is white.[4] Nevertheless, the only things white about this young giant Huc had in tow were the teeth in his smiling mouth. Otherwise, he had the blackest locks and the darkest sunburned skin we had ever seen on a white person. He stood, wordless, before us, a double-barreled shotgun across his back, with both his elbows hooked over it, grinning at the instant attention of our dogs, who jumped around him in delight. The reason for this was clear to us when the giant stripped off his vest, which was conspicuously heavy and bulky in back. He hung it on a tree and, to Kitty's and Holla's continued enthusiasm, hauled out a good dozen squirrels, killed in this one morning. Still wordless, he then pulled out a pocketknife, cut through the base of the tail of the first

---

3. Hauser has Leo using a fascinating German slang term for the police here; he calls them "Polyps."

4. Hauser was translating the man's name for his German audience: "*Chalk* heisst Kreide, und Kreide ist weiss."

animal with a single stroke, and in the next second, with a single tug, had stripped the squirrel of its skin. Huc and I had always needed several minutes for this operation. When all twelve were skinned before our astonished eyes and our dogs had blissfully tussled over the empty strips of skin, he opened his mouth for the first time, to ask a shy question: "Rita, do you think that might be enough for lunch?"

Chalk, as it turned out, came from Grand Tower. He had taken part in the whole Pacific War as a marine and had been wandering all over our farm hunting for a year without our even suspecting his existence until now. "Many times," he related, "I have walked up to within ten paces of you while you were hunting, Henry, but you never saw me."

"Yes, well, why have you never looked in on us, Chalk?"

"Oh, I wanted to see what kind of people you were, first," declared Chalk, with disarming candor. With the same directness, he then disclosed the reason he had approached us now:

"Your boy has the right passion for hunting, but he lacks experience. If it's okay with you, I'll teach him a little bit. Before, whenever I went hunting on the Missouri side, I stayed overnight in those abandoned farms. They all burnt down last summer, and I thought maybe I might be able to sleep in your house, if I could pay my way with whatever I might bag."

With his hunter's and Indian's skills, Chalk was exactly the hero a sixteen-year-old boy craved to admire.[5] Chalk would train not only Huc, but also our dogs, so that an intimate friendship would develop among the four of them. Fitted out with flashlights strapped to their foreheads, the batteries hanging from their belts, with pockets full of cartridges and a couple of candy bars for their only provisions, the men moved out on Saturday evening. Long after that, Rita and I saw beams of light spiriting through the woods on the hillsides.[6] We knew

5. There is a chronological problem here. The year is 1947, and Huc would have been only fourteen, not sixteen.

6. Spotlighting game at night is strictly illegal. Animals transfixed by bright light shining in their eyes typically stand helpless before the hunter. Huc assured me that he and Chalk had never hunted this way. He expressed disappointment that his father emphasized illegal hunting methods when legal ones were just as adventuresome, especially considering that neither he nor Chalk hunted illegally.

that there was little chance that we would see our son and our dogs again until Monday morning. When they returned, almost always heavily loaded with game, the storytelling commenced, lasting for hours. Even the farmwork was interrupted. We heard about the timber wolves that circled their camp all night, whose tracks they saw in the morning; about the fox that Huc had shot right between the eyes in the light of the flashlights; about the old racoon that had outwitted even clever Kitty; about the wildcat's den in a hollow tree. Since the boy had helped me faithfully with the work all week, I gladly granted him his enjoyment with Chalk. Not least of all, because through this friendship he ripened into a man, and I began to feel old, since it now became the case that when we did things together, Huc worked well ahead of me.

It was clear that Rita and I didn't learn about everything that happened on these hunting expeditions. First, only after we had eaten a wild goose supposedly "found" at an unusually early time in the season, did the two confess an unusual hunting method, which was so clever that all we could do was laugh about it. Before the hunting season began, while shooting was still strictly forbidden, Chalk dug a shallow ditch in a suitable place on the riverbank. The ditch deepened as it went land-inward, until it was the height of a goose. It was also built with an overhang, out of which a goose's neck, but not its body, could extend. Chalk baited it with corn, and almost every morning the trap provided him with a goose. The wild geese followed the corn along the ditch until it narrowed to where there was no escape upward or forward. Such a clever animal never arrived at the idea of waddling backwards.

In late summer, as the corn ears hung downward luxuriantly heavy, we got to know in a remarkable way a large number of our neighbors, whose existence until now had been just as hidden from us as Chalk's. It happened this way:

We sat in the yard at supper, and, as usual, Isolde, my favorite goose, squatted on the bench next to me with her bill snapping at the edge of my plate. Then Isolde, with hearing even sharper than the dogs', suddenly began to quack. Usually this meant the present arrival of an automobile, and in a few minutes it came rattling down the hill, the oldest of Henry Ford's products I had ever seen. It died, shuddering, as it came to a stop, a small geyser spewing out of its radiator. The interior of the obviously homemade body was stuffed full with a half dozen pale, mealwormlike women, at least as many children, and twice as many dogs. The only man in the car clambered artfully over a wire that held the front and rear fenders together and upright at the same time. He came running up to us with a small wooden box under his arm.

"God's greetings! The ladies in the car ask if they can perhaps get some water from your pump. And may I perhaps bring you a message from heaven in the meantime?"

I had no time to answer, as the little box was already open, revealing itself to be a portable phonograph, whose crank the pale man immediately began winding with great speed; zip, and a record was clapped onto the turntable, the needle scratched, and after a breathless pause a prophetic voice raised itself:

"The end of the world," it thundered, "is nigh. Daniel has already predicted that Jehovah warned his own ever anew by the examples of Babylon and Nineveh, by Sodom and Gomorrah, and by the fall of Jerusalem, but the heretics have hurried from one deadly sin to another . . ."

The ashen-faced man had crouched behind the Gramophone with his eyes rolled upward, and he listened ecstatically to its words. "Is it not wonderful; is that not God's truth?" he shouted at such authority.

After fire and brimstone had dutifully been rained down upon us and we had swung in Satan's claws over the maw of Hell for a while, the words became milder: There was still the possibility of salvation, certainly not for all of depraved humanity, but only for the small flock (144,000, I believe), who could decide to be Jehovah's Witnesses.

"It's exactly so," stated the white-faced man with a hefty nod and regarded us with a penetrating stare: "Art thou bathed in the blood of the lamb?"

We said no with embarrassment, but the fanatic didn't seem to make much of that. "Well, at least you didn't chase me off your farm with the dogs, like so many other farmers do. That gives me hope that you will permit us to hold a revival meeting on your farm. Some of your neighbors have already said they will come."

As absurd as the outward behavior of this sect appeared to me, under any circumstances I prefer the true believer to the lukewarm. I gladly gave him my permission, even though I knew that in Altenburg and Wittenberg many people would find fault with me for it. With Huc leading, the ancient Ford teetered down the gravel bed of the creek, where, at a place near the river and under a high cottonwood, a campfire was presently burning.

Sharply outlined against the golden cloud banks of sundown and the bloody waves of the Mississippi, a strange and almost biblical column moved rapidly up to this fire. At the front strode an old negro woman with a corncob pipe in her mouth. Following her in a line like a rank of organ pipes were a dozen "pickaninnies," negro children, each with a favorite pet: a hen on her shoulder, a tame squirrel on his kinky-haired head, or a small dog in his arms. Tottering along the railroad bed on canes were a couple of old men from Wittenberg, each led by his only friend, his dog. Hanging on to the apron strings of his tubercular daughter was Negro Bill, now completely blind, of whom it was said that he counted well over a hundred years.[7] From the hills in the north and from the south came the rattletraps of the poorest and oldest farmers, who lived as if invisible in the wilderness ten or fifteen miles away, and even some horse and mule teams were there. It seemed impossible that the Jehovah's Witnesses had sought out all these people. The news had mysteriously spread, as if by bush telegraph, that there was to be a prayer meeting at the "wine farm."

The big gray bundle on the top of the Jehovah-car turned out to be a tent, which was now set up. With grand gestures, the missionary preached in the firelight and prayed that the spirit of Jehovah would be permitted to come to his people. Endless hymns echoed over the river, now turned to silver in the moonlight. Like shadows, the black children ran to and fro with fresh firewood, and the women, unearthly in white dresses sewn from feed sacks, carried ears of corn from our fields for the Eucharist. The religious mood was like that

7. Hauser appears to be introducing a new character here, although he treats him as having an antecedent in the narrative. He has told us at length of the exploits of Leo Harris's erstwhile friend Black Bill, and he has mentioned the "old, half-blind negro, Dick Wilson," as the person probably responsible for the forest fire. The term *pickaninny* has innocent origins; it is probably from the Portuguese *pequenino*, "very little." The term is, however, nowadays considered offensive.

of the book of Exodus; these people had turned away, in revulsion, from the fleshpots of rich America, perhaps because not enough of the riches had been granted to them, and in the twinkling of an eye had fashioned a new life "in the Wilderness." With corn manna that fell from heaven, and with catfish that the negro boys hauled out of the water with the poles they had brought along, the Lord provided for his own.

That night we made the acquaintance of an "America of the Degraded and Insulted," of which most Americans have no concept, a tragic America of people who have not been able to keep pace with their civilization, and who in their backwoods have been frustrated and embittered by it. Stirred by the revelation of St. John, this America, forgotten and overlooked by progress, let its voice be heard, and it was the voice of Job:

"What shall I do?" called out a woman, wringing her hands, "My husband was a traffic policeman, and a couple of years ago a car cut off both of his legs. He took to drink because of pain and grief. He has drunk away his whole pension—the last crust of his children's bread. Sometimes I took away his artificial legs, so that he couldn't get to the tavern, but then he crawled there. Brothers and sisters, he crawled on his elbows in the dust of the highway for miles, until a car picked him up and took him there. Then I was called a cruel woman, but I still had to protect the children, and it breaks my heart, because I know that my husband only drinks because without his legs he has no more life. Such a fine, elegant man he was. Tell me, my brothers and sisters, what shall I do?"

"And what shall I do," murmured old Farmer Weber next to me, "I have 650 acres of land, good land, but my old bones just won't do anymore. I have three strong sons; all three went to war, and all three came home whole and healthy. Two of them have already written that they don't want to take over the farm. They've taken to the city. All through the war I put my hopes in my third son, and I built a road with my two hands, fifteen miles long, so that the boy could get into town easily. The young people today just have to have their movies, and the young women have to go to the beauty parlor. And now the last one, Willy, is gone, too. He doesn't want the loneliness anymore; he'd rather work for wages for the railroad. What should I struggle for? What has been the purpose of my whole life?"

"Be content," said Schrader, also a distant neighbor that I had never seen before, "your boys are at least still alive. I had only one. He was in the woods cutting timber and a falling tree smashed his leg. It was in winter, when the wolves were hungry, and when we found him the next morning, there lay three dead wolves—and what was left of my son."

After midnight, as the full moon sank, the belabored and burdened began to break up, and as the sun rose, the Jehovah-car rolled away to the south, with the sun ball shining through, like an Elias wagon. One could say that the whole thing disappeared like a ghost, except that it was no ghost; rather, it was the tragic reality of the decline of the old America.

The harvest this year was splendid. At two dollars and ten cents a bushel, the corn price had reached a high seldom ever seen, and hog prices were up to twenty-eight and sometimes thirty-one cents a pound, live weight. We made about a third again more on both corn and hogs than we had the previous year. The last days with our boy were spent happily forging plans: We wanted to build a proper barn the next year, and three miles of fence for more livestock. Above all else, though, Huc wanted to study electricity over the winter, in preparation for our boldest plan: an electric plant for our farm. We wanted to buy the big stern wheel from the *Golden Eagle,* and bring it to our side of the Mississippi. Built into the river current, it would serve as a turbine, a power source for a generator. A wire would be brought under the railroad bridge and from there up to the house. Electric water pumps and electric lights, an electric refrigerator and brooder, an electric sewing machine for Rita, and running water in the house. What didn't we have in mind!

The Black Year

On New Year's Eve we stepped out onto our veranda as was our old tradition and stood hand in hand, until we could hear the bell sounding all the way over in Grand Tower. I handed the shotgun to Rita, and she fired a shot over our dark fields to greet the year 1948, the only shot she ever fired, from New Year to New Year.

Then we lit the Christmas tree for the last time, and in its light we tallied up the old year and planned in advance for the new one. The result was not so favorable as Huc and I, in view of our good harvest, had estimated. Despite all of Rita's efforts, the Angoras had lost money. After two good harvests, our ground had to be replenished with what had been taken from it. That meant it had to be richly fertilized, and this meant in turn that no thought could be given to a new barn and three miles of fence, at least for the time being. Our tractor, as well as Perfidio, was ripe for an overhaul, and a proper tractor-pulled plow and seeder were overdue. The house cried out for new paint and for termite protection. We hadn't bought any new clothes or underwear in years. The greatest and leakiest hole in our economic plan, however, was Europe. Over a hundred packages had been sent to Germany in the last year and there should be at least as many in the new year. The costs for this had come to thousands, and suddenly it was shockingly clear to us that in no way had all this outlay been paid for by our farm, but that we had stripped our land as well.

A dark shadow fell over the new year, therefore, and this was strengthened by an event that is hard for me to relate, because one might perhaps not understand how incisive this was for us.

In January, led by a snowstorm, a cold wave quite unfamiliar in southern Missouri moved over our land. Our only connection with the village was by foot along the railroad embankment, where the snowplows at least kept the rails open. With the dull thunder of Mississippi ice grinding in my ears, I set out one morning for Wittenberg, to get the mail and to make some small purchases. As always, I had to constrain the dogs sharply, to keep them from following me. Accustomed only to life in the forest, sometimes they were grasped by panic when a train approached. I also didn't want

154

them to get too friendly with the village dogs. Leaning against the hard wind, constantly slipping off the icy rails, my eyes slitted against the swirling snow, I already had three quarters of the way home behind me when from the south I heard the locomotive of the 11:40 passenger train howling. Annoyed, I stamped off to the right into the deep snowdrift left by the plow, until I was buried up to the hips, blinking toward the oncoming train. Again and again the whistle howled eerily, as if the engineer wanted to warn someone. Perhaps he had seen me, I thought, but I was not at all on the right-of-way and in no danger. The ground began to shake under the momentum of the mass of steel. For several seconds I felt as if I could see something like a snowball racing through the whirling flakes in front of the machine. Then the jet of steam from the cylinder hissed away over me and the undercarriage of the long express train cars thundered by, until in their vacuum a lashing snow whirlwind stormed over me.

With a curse I worked my way out of the snowdrift and would have jogged on ahead, but ten paces ahead I spotted a large, fiery red spot. Next to it, barely distinguishable from the snow, lay a body, a dog—my Kitty, with her head crushed in. Her pink tongue still moved, her eyelids and her legs twitched. Steam rose from the blood of her wound, and then she was dead.

Only people who live in great isolation, only people who live so intimately with animals as we did, can understand that I threw myself over her body with a cry; then I pitched all my purchases out of the rucksack, so I could carry my Kitty home, her heavy, soft body radiating its last warmth into my back. Yes, that was harsh, but what came next was the worst. Then as I brought Rita the message, she simply stared at me, stunned, pale as death:

"Kitty? Oh, God, she wanted to get you to come and help!"

Sobbing, she threw her hands over her face, and erratically, the words came:

"This morning I wanted—to fetch the eggs.—The steps were icy—I slipped and fell—hit the back of my head—unconscious.—I woke up, and there was Kitty, licking my face.—She left me no peace until—I was back on my feet—in the warm house.—I would surely have frozen—if Kitty hadn't found me.—And then she ran off to get you.—She saved my life, our Kitty—and now she has died for me."

Gripped in wild despair, I cried out, "I can't bear this any longer. We have to do something. Rita, come. We have to bury our Kitty. And since she died for us, we'll make her bed in our cemetery—right next to the graves of the children. I doubt that it will displease the children or that it would be a sacrilege."

It took a long time until I had loosened the hard-frozen earth with a pick enough to be able to dig, and I am not ashamed to say that we cried as the little grave was filled and Kitty's sister Holla sniffed at the gravestone with sad, questioning eyes.

In spring it appeared as if the year would be a good one. We dealt with everything intensely in order to make it all as nice as possible for the arrival of our son. The heavy five-gallon pails with white lead paint for the house were already arrayed on the veranda. And then it was as if a veil, in great and small things, draped itself over our happiness. At the beginning of May, a late spring flood washed out my already seeded fifteen acre field. By the time the field dried out, it was too late to replant. A little later, a twenty-hour cloudburst fell, with such force as experienced here only in each generation. It tore out no fewer than seven timber bridges on the road to our farm. In the streambed, which had become a rushing river, lay thousands of tons of gravel and stones piled high, which had also widely covered the edges of the fields. On a reconnaissance trip through the devastated terrain, I spotted hidden tree trunks in the high grass. The sudden yank of an impact was followed by the ominous sound of onrushing water, and I discovered that Perfidio's radiator was split from top to bottom. Replacement parts for such an old car normally could not be found. After much correspondence, an auto wrecker in St. Louis delivered a halfway suitable substitute for a lot of money, but the car was laid up for months. The people in Europe wrote of sickness and ever-climbing poverty, and our bank account had once again sunk below three figures. Even Leo seemed to have been pursued by the streak of bad luck. "Now I can go fishing again," he reported one day, sadly. "The cops found my still."

"How did that happen, Leo? You said it was well hidden."

"It was, too," he grumbled, defiantly. "Only I hadn't figured on the damned wild geese."

"What did the geese have to do with it?"

"Yeah, well, you see, Henry, it came to this: The geese ate the mash

I had thrown out and got drunk on it. A watchboat was patrolling up and down the island, and the cops wondered why the geese on the shore didn't fly away and were staggering around so peculiarly. They tied up, and simply followed the goose trail backward. Then nothing else could have happened but that they stuck their noses right into my copper boiler. There were sixty gallons of booze in it, almost ready. A real pity. Well, at least they didn't catch me."

Our son came, this time burdened with a miniature stretcher, between whose poles mysterious metal boxes were built, and out of which hung wires. This, he explained, was an American army mine seeker, on sale after the war for a dime a dozen.

"For heaven's sake, what do you want with it?"

"Don't you remember, Father, how the town people talked about the gold treasure that had been buried on our farm during the Civil War? This device will find anything metal, down to eighteen feet deep. I want to make us rich, Father, and even if the gold treasure is only a legend, lead has been found in our hills and maybe I'll find uranium or other valuable metals."

Treasure hunting with the help of old mine seeking gear was in full swing among American boys at that time, and even though Huc's expeditions into the hills brought nothing more to light than the fool's gold of pyrite, he declared fixed and firm that great underground wealth existed. In a deep mudhole in the streambed, whose surface shimmered with oil, he presumed he had found an oil well. In fact, a few miles south of us on the Illinois side, petroleum had been found recently. His eagerness to make us "rich" moved me all the more, since the boy well knew how poor we were in reality, and that this summer we would become even poorer by the day. Be it that the hunger of the world had lessened, be it that the speculators in the Chicago grain exchange had overbought, but corn prices fell in an alarming way, from $2.10 per bushel in January, to $1.10 in August. For us, since we had purchased tractor equipment in the spring on a bank loan for almost a thousand dollars, the corn price drop of nearly a half meant an economic catastrophe.[1]

1. The economic catastrophe that Hauser mentioned here remains catastrophic. Spot market corn prices at harvesttime in the 1990s remained as

So it was then that with a heavy heart I began to paint our house in August. The longest, straightest young maple trunks from our woods were made into ladders. With wire gauze masks on our faces and DDT bombs in our pockets, Huc and I climbed up to the roof first to exterminate the countless wasps' nests, which hung everywhere from the edges of the old boards. Then followed the boring scraping of the old paint with wire brushes and putty knives, the ground coat of linseed oil, and finally the shining ivory of the white lead paint. Every time we admired our day's work at supper in the garden, and Huc declared, "No house in the whole world is as beautiful as our house," I felt a stab through my heart. Why did we do it? Was it really for us, or was it for the bank, who would doubtless have to force the sale when farming in our area, with us ahead of all, became insolvent? In a strange manner that couldn't be explained by the return of an old bout with malaria alone, my bodily strength and energy diminished that summer. A premonition drove me one day to go to the county seat in Perryville and legally transfer the farm to Huc's name. Whatever happened to Rita and me, the farm had to stay with the boy.

At the beginning of September, when transporting the corn to Altenburg began, Huc returned with the empty wagon increasingly discouraged: "Fischer only wanted to pay ninety cents a bushel. Father, this isn't worth it any more." It cost us great emotional effort to disclose to the boy our own discouragement. It was his summer, perhaps the last summer with us, and we wanted it to be a pleasant summer . . .

In Indian summer, when the sumacs painted themselves red, we began with the repair to the bridges on our road, for which there had been no time in the summer because of undisputed work. Huc dragged with the tractor big cedar trunks, which would serve as lengthwise beams, while I dug the supporting abutments and lifted the beams into place with jacks. Over them we nailed a layer of bridging, which Huc covered over last with a layer of gravel. Handsome our bridges were not, but capable of supporting even

---

low as he experienced them in 1947. Overproduction has kept agricultural produce market prices low for decades.

the weight of a heavy truck. For improvement of our fearfully and abundantly rutted road, Huc invented a brand new method, that in Europe would have been considered wasteful, but which for us was completely natural. At the abandoned sawmills there lay everywhere, in giant stacks, the slash or side boards. We filled the ruts, often axle-deep, with them, and then we covered them over with thin layers of gravel and sand. This year I no longer felt it so disgraceful that my son "passed me by" in his work. Occasionally, while I waited on the tractor with a fresh load of boards, I sat at the edge of the road with a darkened disposition; "Why are we improving our road now? Only for the people who will haul away everything movable that we have made or bought for our farm?"

The fateful letters from Europe arrived as the wild geese were beginning to migrate, on one of the last days that Huc was still with us. The two most beloved older people that Rita and I had in Germany were both so ill that they very well might not live through this year. Almost with the same words both wrote, "Can't you come, so that I can see you once again?" Wordlessly, Rita and I exchanged our letters. Silent and with trembling lips, we looked at each other afterwards, in our eyes reading the shared decision: "Europe—after all, yes, it's time."

That evening we discussed things with Huc, not as if he were an underage son, but as a partner, a man. With his big, still soft, yet hard-working hands on his knees, his dark head lowered, he sat there for a long time without speaking. Then, with a furrowed brow, he said, "I think it's right that you should go. Father, you are a writer, and German is your mother tongue. And you, Rita, are the eldest of your family, and they will need you especially in the ruined Old Country, much more than here. You both probably really must go. Now I want to go out. I want to take a long walk in the moonlight; I want to be all alone on our—my farm. Don't wait up for me when it's time to go to bed."

The next day, as Perfidio, with Huc at the wheel for the last time, climbed the steep hill behind the house to go to the station, the boy pulled on the brake at the highest point. Without looking at us, rather with a last long look over our house and the river, he said, fighting back tears, "You have to sell everything that's movable; otherwise, it will be stolen. Just promise me one thing—that you won't sell Holla. Give her to Chalk; he'll take good care of her. And, okay, my rifle. I'll keep that . . ."[2]

Never had our farewell been so leaden and sorrowful as this time.[3]

Not long after, I traveled to Washington to discuss our return to Germany with the officials. That was done quickly enough. After we had fought for our citizenship rights for nearly ten years, I got our exit permission in a shake of the head. "Now the die is cast," I said to Rita on the way home. "There is only one difficult thing yet to live through—the auction."

With Mr. O'Connor, a former sheriff, who was known far and wide as a good auctioneer, we discussed the date. He recommended the eleventh of November, a Saturday, on which no other auctions had so far been announced. Then we took a list of our possessions to the print shop of the *Perryville Gazette,* which delivered the handbills to us by mail. They were big and red; at the head, as the chief attraction, was the tractor, and at the end, "other items, too numerous to mention." The mail truck driver, Roth, divided the fliers among the villages between Perryville and Cape Girardeau. Leo Harris took a second packet over to Grand Tower. I had to deliver them to Wittenberg and Altenburg myself. It was bitter, there, to go into the bank and the stores. "May I please post my auction notice in your window?" Our old friends allowed it, dejectedly, with averted looks, and in the eyes of others there appeared an absent expression, as if I really was no longer there.

In the evening of the tenth of November, a heavy cloud bank came over us and threatened a persistent rain. "Do you think the weather will hold?" I asked again and again, as we hauled machinery and household goods onto the veranda.

Without interrupting her melancholy labors, Rita shook her head, "We have no power over the weather; what comes, comes."

---

2. This is a bit of poetic license. The house is not visible from the gate, and the road to Wittenberg in fact climbs well out of sight of the farm's lands and structures.

3. This departure was sad enough, but its poignancy is intensified when we consider that neither Hauser nor Rita ever saw Huc again.

"If a heavy rain falls today, our road will be impassable. Then nobody will come to our auction, or only so few that we wouldn't even be able to pay our ships' passage."

"Dear, it's pointless to torture yourself with these thoughts."

"I know it, but there is one emergency escape. If we take our things to Wittenberg, then the auction can be held there at the post office. The worst stretch is the rocky hill right behind our house. I want to load the wagon with at least the most important machinery and haul it with the tractor at least to the hilltop."

Tired and sorrowful, Rita shrugged her shoulders, "Whatever you think."

It was an act of desperation. I myself didn't believe it could help us in any way. It just allowed muscle exercise and the opportunity for many a bitter oath. At least ten times that night I got up to look at the sky, which appeared to sink lower and lower, starless and threatening. Each time, when I returned to the cots that, as in the first of our time here, had become our beds, Rita grasped my hand, startled out of a nightmare; "How is it outside?"

"No change." So passed this uneasy night.

It dawned a misty, gray day. Even before coffee I drove the tractor back to the farm. The house, surrounded by all the things that belonged inside it, the yard with the fishnets, hauled out of the river and now hung in the trees, this all seemed so unfamiliar. I couldn't endure it, least of all the sight of my wife, polishing the washtubs mirror bright for "the others." There was still another hour at least, and nothing more for me to do after I had fed the animals for the last time. Then I ran up Cemetery Hill and knelt between the graves, where no one could see me, my head buried in my hands.

I cannot and do not want to describe what these hours were like. Ten years of American life drifted over me: New York and our first basement apartment in Greenwich Village, being able to breathe again in freedom, but this mixed with fear for our life and a feeling of dreadful desolation. The first farm in South Valley, sixty miles west of Albany; wrestling with rocky soil for five years, in happiness and sorrow. The dreadful winters with Siberian cold, in which, at temperatures forty degrees below zero, whole tree trunks in the woods burst with the thunder of cannon. Working as a lumberman in three-foot-deep snowdrifts, which broke my health—and Rita, paid by the piece for sewing gloves, which kept us alive through the war. The cruel streets of Chicago, and San Francisco's parks, in eternal springtime and enveloped in the soft thunder of the Pacific surf. I thought about seven years of struggling with the new language and about my acceptance into American literature. There was the growth and maturing of my children, and the pain over the one that was caught up in the labyrinth of dispute and error. And now, finally, the farm; our "second birth" in the new land, which we experienced when the last cells of our bodies born in Europe died, and the severed old roots grew again and drew energy from what was now our mother soil. And then the stranglehold of the survivors in the Old Country, who, without knowing or wanting it, twisted around us like vines, strangling our new life and sucking the pith from the revitalized stems of our lives. Would it be possible to love the Old Country as we once did? Filled with shame and grief, I felt the love that had burst forth again—expressly with the collapse of the Old World—mixed peculiarly with denial, indeed, virtually with hate. Together they had worn down the core of our lives: an America we could not resist, and a Germany that had become a vampire. Yes, and now here was the defeat, our entry into the no-man's-land of "neither here nor there." Transplanted twice, the old tree could not sink its roots again.

It had got to nine o'clock, and from the hills the first of the farmers' vehicles rattled near to view the goods on exhibition, an hour before the auction. I felt it my duty to play the roles of host and lord of the manor, which I no longer was. I noted very quickly, however, that this was quite superfluous. These people were strangers, and since auctions in these parts are like festivals, they were perfectly at home. As was their perfect right, the men tried out the tools, started the tractor, and estimated the swine with cool, calculating eyes, while their wives wandered through the house, curiously and somewhat derogatorily handling every article, and with always the same questions to my wife: "So, you people are going back to Europe? Well, aren't you a little afraid? Do you think it's going to be better there than here? I'll bet they're going hungry over there."

No, I was superfluous until the auctioneer came. I ran down the road to Wittenberg to a place where I could hide in the bushes and count the cars, this being a standard for the success of the sale. Up to ten o'clock I counted twenty-five cars. That wasn't particularly good, but not particularly bad. Women came from the village, too, to take over the care of the crowd, with big coffee pots, huge containers of frankfurters, and baskets of white bread. But where was the head man, the auctioneer? Mr. O'Connor had promised firmly he would be punctual, and now it was already a quarter of an hour past time. Three, four more cars dribbled in, with those genteel folks who didn't want to concern themselves with the junk with which every auction began. Then finally, finally, the great man arrived—O'Connor and his two sons.

I jumped on the running board of his car and immediately received my instructions: "You, Henry, stay near me and haul up the things I'll be bidding up. Your wife will play scribe and write down the sales; tell her to pay close attention. Fred here, my eldest, will be cashier, and Gale, my youngest, backs me up when I get hoarse. Some people say he's a better auctioneer than I am myself. How many people, do you reckon, are here? Maybe a hundred and fifty? Well, that's not particularly good, but not bad, either. The main thing is that the sun comes out. Then the people will be warm. Have you heated the house? Good—the women can warm themselves there."

Fully conscious of his own significance, indeed, almost majestically, the stocky man with the sharp-edged, shrewd face strode up the steps to the terrace. The crowd pressed forward, waiting avidly until O'Connor stamped ceremoniously on the boards three times with his white cane.

"Folks," he began, with a deep, pleasant voice, "you've all heard and read that Henry, here, and his wife, are giving up their farm and going back to Europe. I have looked over Henry's things. These are good possessions, in large part almost new. Henry, so I hear from all sides, has been a good neighbor to you. So, don't ponder too long on your offers; let's carry out the auction smoothly, so we won't get back home too late. You'll help Henry to get home quickly that way, and above all, you'll help yourselves, because you'll never have an opportunity like this again. Okay, let's get to work. Henry, hand me that basket with the canning jars. I have here three dozen quart jars, as good as new. These days in any store you'll pay a dollar twenty a dozen. Who'll bid the first dollar for the three dozen jars; that's a gift, my friends, one dollar—one dollar—one dollar! Nobody?

Who'll make an offer, any offer, twenty-five cents? I have an offer of twenty-five cents." He raised his arm and pointed with his stick at someone who had slowly raised his hand. "Thirty—thirty-five—forty—fifty cents! Nobody else? Really, nobody else? Fifty-five cents. Mrs. Homann has bought the basketful of jars for fifty-five cents."

Deliberately begun deeply, his voice had spiraled upward like an accelerating dynamo to a high singsong. A kind of numbing effect resulted from it, a fever that the people caught, and that even I yielded to, like a very sick person to the anesthesia. As if in a dream, I dragged the things up, and like dreams I saw them disappear. The picks, the shovels, the axes, the saws, the cooking pots, the pans, the mattresses, the stove. I knew what every one of these things had cost us in hard work—many memories hung on every one of them—and with every one of them pain jolted through me as it was squandered. And yet everything was made unreal by the dervish dance of the auctioneer's voice. I could bear what an hour ago had seemed unbearable, and somehow I was thankful to the people who carried off our things with the compulsion of a column of ants. It was as if a grave were being filled, shovel by shovel.

About noon, when it came to the auctioning of the livestock, old O'Connor let his son replace him. The younger man's stress was harder and sharper, his eyes more sparkling, but his voice was to the hair the same, the uninterrupted suggestion and increase in the moment of tension. Trucks rolled backward up to the enclosure, dozens of men jumped over the fence to round up the animals, and I walked behind a tree and held my ear against it, so as not to see, not to hear, the outraged and pathetic squealing. Later, when it was the Angoras' turn, they went for fifty cents apiece. I saw Rita standing on the edge of the ring of buyers, privately wiping her eyes, yet with her jaws clenched she got through this weak moment, just as we got through it all.

In the afternoon I found out how a good auctioneer was even able to bully a listless crowd. It was Perfidio's turn. In our hands alone the trusty old car, dented and scratched, had traveled sixty thousand miles of country roads, heavily loaded. Like a dinosaur out of auto-prehistory, he stood among all the modern cars, and nobody wanted him. In vain old O'Connor had stamped threateningly with his cane: "No offers? No offers at all, not even a hundred and fifty dollars?

Then I want to tell you something: I myself hereby purchase Henry's car—for two hundred dollars. Is that all right, Henry?"[4] The crowd grasped the lesson with discernment, and the offers for the farm machinery were immediately much livelier. The tractor, which as the best horse in the stall had been held for the last in order to bring the crowd up to scratch, brought about a tough battle. At $500, the professional buyers, who were only interested in good opportunities, had already dropped out. At $650, two out of the four seriously interested bidders pulled out. At $700, the two who were squabbling over ownership like fighting cocks sat down opposite one another, and half seriously and half in jest, began to pull one another back and forth by the hand. Like a referee in the ring, O'Connor enjoyed the scene and drove the winner up to $750. That was more than I myself had paid. Then house and yard droned with the starting of many motors. Gears ground, the babble of voices ebbed away, and after a few minutes it was quiet in the yard, eerily quiet and empty; there were no more animals.

In the kitchen, which throbbed with emptiness, sat the three O'Connors, totaling up. In all, $1,711 had been taken in, minus the O'Connors' fee of $25. After paying off the bank, this would just cover the ship's passage, the train trip to New York, and leave perhaps a hundred dollars as backup funds.

"Can I go with you to Altenburg?" I asked the old sheriff. "I want to go to the bank and then deal with some other things."

He nodded, "Of course. I know pretty well how it is for you, Henry. I'll drive you back to Wittenberg, too. You don't have a car anymore."

Later, as I walked back along the railroad embankment, it was already almost night. On the steps to the terrace sat Rita's lonesome shape, almost rigid, her eyes fixed on the last glow of the sunset. She didn't notice me.

---

4. Despite Hauser's published account, Henry Scholl remembers the sale and the auctioneers, and his recollection is that the auctioneer was a man named Ed Nations. He is sure of this because Mr. Nations is the individual who bought "Perfidio." Mr. Scholl was asked on occasion to drive the old Packard to a garage for service, and he recalls that it was the property of Ed Nations at the time. It is not at all clear why Hauser, otherwise so accurate in his narrative, often forgot or changed names.

Mute, I sat next to her, laid my rucksack between my feet, and let my eyes sweep over the harvested fields, which had served as parking lots and were now rutted by many tires. Our Holla ran around them like a lost ghost, whimpering restlessly and snuffling at the ground. The dog didn't grasp any of this, but she grasped it not much less than we.

"Give it to me," said Rita, finally, and her voice sounded as eerie and shadowy as her appearance in the quickly fallen night.

"What?"

"What you brought with you."

I opened the rucksack and pulled out the bottle of whiskey.

"I knew it," she said. "Do you know, back then in the first year, when I ran the fishhook through my hand, and you had to cut it off so you could push the barb through?—I needed whiskey then, too, but today I need a lot more."

We drank. In the threatening, dark sky there was no star of hope.

It was still ten days till the ship's departure. We spent eight of those days on the farm, to save money, and those days were of an unearthliness that is indescribable.

We spoke only of the most necessary and most commonplace things. To touch on reality hurt too much.

The house, the valley, the river, they were there as they had been when we came. But we were not as we had been.

The tidy farmstead with its eleven outbuildings, which we had erected, the fruit trees at the edge of the road, which we had planted, the fields, the chief work of our hands, were still there. But we were really no longer there.

Chalk came on the third day and took Holla away. After some hesitancy and misgivings, she followed him to his boat, wagging and jumping, lured by the odor of his hunting vest. Then it was even more quiet around us.

On the fifth day, Leo Harris and Naomi came. The two loaded their boat with the preserves from our cellar. Leo promised us the blue of the sky: He would mow the weeds and plow in the spring, so that no forest fires could reach our house. We thanked him and didn't believe a word. His spirit was willing, but his old flesh was weak. He only talked that way to get away as fast as possible, because it was unnatural for him to be with us now.

On the sixth day, Rita started to pack: The old suitcases from Europe with the colorful stickers from long-bombed-out hotels, and cabin numbers of long-sunk ships still glued to them. That evening a rented car took our things to the station.

On the seventh day, Rita said, "Let's build a big fire and burn up anything in the house that can still burn. Even the food that's left over that no one wants—the stuff will only spoil." I did it, and as the pyre burned, the beans from our winter stockpile crackled, and the pickled eggs hissed, I broke out in crazy laughter and couldn't stop, until Rita shook me by the shoulder with all her might. "What is wrong with you?"

"Nothing," I said, awakening, "But here we are, going to a starving land, and we're burning up all the food we so laboriously gathered. If this isn't insanity . . ."

Early in the morning of the eighth day, we hung our cots and the bundles with our work clothes on lines from the ceiling, so that the rats couldn't reach them. Then we carried our light bags out and locked our house. It was only a gesture, as if no thief could get in simply by tossing a stone through a window.

We climbed Cemetery Hill for the last time. Once we had painted a picture of how nice it would be to rest there. Now there was only a tiny cypress twig from Kitty's grave left for us, which I put in my briefcase.

Then we wandered hand in hand to the railroad bed, to the place where the view of our farm opened. But I saw it no longer. The white house blurred before my eyes. Instead, the gray and ghostly house with the fallen roof that I had encountered on the island emerged. It would be that way soon, our house—soon, too soon. The wilderness closed in quickly . . .

At our property line on the railbed leaned the sticks we used to help us walk on the rails. As I gripped mine, the bark, softened by the fall rain, loosened and fell away. Each of us walking on a rail, I with the white stick in my hand, we made our way to the station, the two of us, silent, without a look back.

Only our hearts cried, "We loved you, and we will always love you—our farm."